**DEPARTMENT OF THE ARMY**
**U.S. Army Corps of Engineers**
**Washington, DC 20314-1000**

CECW-ET

Manual
No. 1110-2-501

1 February 1999

## Design, Construction, and Operation
## SMALL WASTEWATER SYSTEMS

**1. Purpose.** This manual is intended to provide guidance and criteria for the design and selection of small-scale wastewater treatment facilities. It provides both the information necessary to select, size, and design such wastewater treatment unit processes, and guidance to generally available and accepted references for such information. For the purpose of this manual, small-scale wastewater treatment systems are those with average daily design flows less than 379 000 liters per day (L/d) or 100,000 gallons per day (gal/d), including septic tanks for flows less than 18 900 L/d (5000 gal/d), small prefabricated or package plants for flows between 18 900 L/d (5000 gal/d) and 190 000 L/d (50,000 gal/d), and larger prefabricated treatment systems with capacities of no more than 379 000 L/d (100,000 gal/d).

**2. Applicability.** This manual applies to all HQUSACE Commands having responsibility for civil works projects.

**3. Distribution Statement.** Approved for public release; distribution is unlimited.

FOR THE COMMANDER:

ALBERT J. GENETTI, JR.
Major General, USA
Chief of Staff

---

This manual supersedes EM 1110-2-501, Part 1, dated 29 September 1978; Part 2, dated 30 September 1980; and Part 3, dated 31 May 1983.

| DEPARTMENT OF THE ARMY | EM 1110-2-501 |
| U.S. Army Corps of Engineers | |
| Washington, DC 20314-1000 | |

CECW-ET

Manual
No. 1110-2-501

1 February 1999

# Design, Construction, and Operation
# SMALL WASTEWATER SYSTEMS

## Table of Contents

## Chapter 10
## Sludge Disposal

## Appendix A
## References

## Appendix B
## States with Regulations/Requirements Applicable to Small-Scale Wastewater Treatment Facilities

## Appendix C
## Wastewater Characterization Data

## Appendix D
## Wastewater Design Criteria and Examples Matrix Summary from Non-Military Sources

## Appendix E
## Design Examples

**Appendix F**
**U.S. Army Experience with Natural Wastewater Treatment Systems**

**Appendix G**
**Abbreviations and Glossary of Terms**

# Chapter 1
# Introduction

## 1-1. Purpose

This manual is intended to provide guidance and criteria for the design and selection of small-scale wastewater treatment facilities. It provides both the information necessary to select, size, and design such wastewater treatment unit processes, and guidance to generally available and accepted references for such information. For the purpose of this manual, small-scale wastewater treatment systems are those with average daily design flows less than 379 000 liters per day (L/d) or 100,000 gallons per day (gal/d), including septic tanks for flows less than 18 900 L/d (5000 gal/d), small prefabricated or package plants for flows between 18 900 L/d (5000 gal/d) and 190 000 L/d (50,000 gal/d), and larger prefabricated treatment systems with capacities of no more than 379 000 L/d (100,000 gal/d).

## 1-2. Applicability

This manual applies to all HQUSACE Commands having responsibility for civil works projects.

## 1-3. References

Required and related publications are listed in Appendix A.

## 1-4. Distribution Statement

Approved for public release; distribution is unlimited.

## 1-5. Laws and Regulations

*a. General.* The design, construction, and operation of wastewater treatment facilities that either discharge wastewater to surface waters or use natural systems as a disposal method are controlled by Federal, state, and local laws and regulations. The National Pollutant Discharge Elimination System (NPDES) permitting program under the Clean Water Act (CWA) is designed to control wastewater discharges to surface waters. For more details on the laws and regulations governing wastewater discharges, see TM 5-814-8.

*b. Army policy.* Army policy is to use regional or municipal water supply and wastewater collection and treatment systems, when economically feasible, rather than construct or operate Army water supply and wastewater systems (AR 200-1, Chapter 2-8).

*c. State regulations.* Table B-1 presents a comprehensive list of state regulatory contacts. A summary of states with regulations regarding land applications for subsurface disposal of wastewater is provided in Table B-2. Table B-3 identifies the states that have developed specific design criteria for wastewater treatment systems. Chapter 10 presents a detailed discussion of applicable Federal sludge disposal regulations, 40 Code of Federal Regulations (CFR) 503.

*d. NPDES program.*

(1) The NPDES permit process is authorized by Section 402(a)(1) of the CWA. Under the NPDES program, each operator or owner of a wastewater treatment facility desiring to discharge wastewater to surface waters (lakes, rivers, creeks, oceans, etc.) is required to obtain a permit for such activity. The authority to issue permits may be delegated to states meeting certain technical, administrative, and legal requirements. The NPDES program is administered by ten Environmental Protection Agency regions and 35 approved NPDES states as of January 1, 1994 (see Table B-1). The CWA does not preclude state or local authorities from promulgating more stringent standards than those required under the national standards.

(2) The NPDES program in its current form has evolved from a number of legislative initiatives dating back to the mid-1960s. The amendments to the 1972 legislation (Clean Water Act of 1977 and Water Quality Act of 1987) shifted emphasis from controlling conventional pollutants ($BOD_5$ and TSS) to controlling toxic discharges.

(3) NPDES program authority can be divided into four elements: Municipal and Industrial Permit Program; Federal Facilities Program; Pretreatment Program; and General Permit Program.

(4) The authority to administer the NPDES program to Federal facilities is a programmatic responsibility assigned to NPDES states and also covers any facility that discharges less than 379 000 L/d (100,000 gal/d) of wastewater. Table B-2 identifies the states with NPDES program authority.

In those states where the NPDES permitting authority has not been delegated, the facility will require a state and a Federal permit. In addition, the CWA (Section 313(b)(2)) added a significant requirement for Federal facilities constructed after September 30, 1979, to evaluate innovative wastewater treatment alternatives. Recycle, reuse, and land treatment technologies are considered as innovative. According to Section 313, innovative technologies must be used unless the life cycle cost of the innovative system exceeds that of the next most cost effective alternative by 15 percent. However, the EPA Administrator has the authority to waive this requirement.

*e. Pretreatment program.*

(1) The pretreatment program was developed to control discharges to Publicly Owned Treatment Works (POTWs) or those that have the potential to contaminate sewage sludge. The pretreatment program establishes responsibilities of Federal, state, and local government, industry, and the public to implement National Pretreatment Standards to control pollutants which pass through, or interfere with, treatment processes in POTWs or which may contaminate sewage sludge. The regulations developed under the pretreatment program apply to pollutants from non-domestic sources which are indirectly discharged, transported by truck or rail, or otherwise introduced into POTWs.

(2) The term "pretreatment," as defined in Part 403 of the CWA, means the reduction of the amount of pollutants, the elimination of pollutants, or the alteration of the nature of pollutant properties in wastewater prior to introducing such pollutants into a POTW. The reduction, elimination, or alteration may be accomplished by physical, chemical, or biological processes, process changes, or other means, except as prohibited by CWA (Section 403.6(d)). A more detailed discussion on the effects of toxic pollutants on biological treatment processes can be found in TM 5-814-3, Chapter 3.

*f. Effluent limitations.* The NPDES permit effluent limitations are developed in each site-specific case by three methods: effluent limitations guidelines; water quality considerations; and best professional judgement (BPJ). In general, effluent limitations guidelines are employed in cases where water quality standards are not contravened. Such limitations are technology-based and represent "end-of-pipe" technology. However, the owner or operator of a treatment facility can use any technology that achieves the same effluent quality standards. Many situations require the development of limitations based on water quality considerations. Usually, water-quality based limits are required only for selected parameters which are shown to be toxic to the aquatic environment. BPJ is used in cases where effluent limitations guidelines are not available for a particular pollutant parameter.

# Chapter 2
# Preliminary Data Requirements

## 2-1. General

The goal of Federal and state water pollution control authorities in conducting pollution abatement activities is to protect and enhance the capacity of water resources to serve the widest possible range of human needs. Material presented in this chapter is intended to identify the data requirements considered necessary to the design of small-scale wastewater treatment facilities.

## 2-2. Recreational Facilities

   *a.* *Definitions.* The term "recreational area" is used throughout this manual to include land or water areas dedicated to the enjoyment of the public. For the purpose of this manual, recreational treatment facilities are defined as any wastewater treatment facilities for recreational areas including primitive campsites; modern campsites complete with trailer dump stations, flush toilets, and showers; and parks, picnic areas, overlooks, comfort stations, fish cleaning stations, etc.

   *b.* *Type.* The type of recreation area determines the complexity of the recreational facility treatment system. For example, a modern campsite requires a more complex design of the wastewater treatment facility than a primitive campsite, while recreational treatment facilities in parks, picnic areas, overlooks, comfort stations, and fish cleaning stations have special design considerations of their own.

   *c.* *Frequency.* The frequency of public visitation is an important consideration in the design of any recreational treatment facility. Most recreational treatment facilities are seasonal operations and experience wide fluctuations in wastewater flow that can range from no flow to maximum flow conditions over a short period of time. For example, facilities that experience large number of visitors on weekends may require a treatment process that can effectively operate over a wide fluctuation of both hydraulic and organic loading.

   *d.* *Estimation of design parameters.* The estimation of wastewater design parameters has been historically based on different methods, such as traffic count, percent occupancy, and head count. Each method, however, has inherent limitations and may or may not be applicable to a specific site. A detailed discussion of each estimation method is presented in Chapter 3.

## 2-3. Determination of Effluent Limitations

   *a.* *Regulations.* The primary design goal for any wastewater treatment plant is to meet Federal, state and local effluent limitations and receiving-body-of-water quality standards. Therefore, the design engineer must become familiar with national and local regulatory requirements governing a specific area for discharging wastewater and/or land application.

   *b.* *Monitoring requirements.* Federal and state regulatory requirements for discharges from treatment facilities into recreational waters are usually more stringent than those for discharges from treatment facilities to other receiving waters. Monitoring requirements usually consist of flow, residual chlorine, pH, 5-day biochemical oxygen demand ($BOD_5$ ), total suspended solids (TSS), and fecal coliform. Total Kjeldahl nitrogen (TKN) and total phosphorous determinations may also be required. Table B-3 summarizes state requirements pertinent to recreational treatment facilities design.

## 2-4. Site Selection Factors

*a. General considerations.* The planning design engineer, when selecting sites for recreational treatment facilities, must ensure that the planned facility will not cause interference or detractions from the natural, scenic, aesthetic, scientific, or historical value of the area. In addition, topographic, geological, hydrogeologic, and atmospheric factors and conditions must be considered when designing the treatment facility for a recreational area. For specific considerations regarding site selection, space, and access requirements, see TM 5-814-3, Chapter 2.

*b. Aesthetic considerations.* The designer must ensure that distinguishing features that make the area of recreational value are not degraded. Vertical building construction should complement or enhance adjacent architectural and environmental features. Aesthetic aspects are important enough to the value of any recreational area that additional construction, operation, and maintenance costs to preserve the beauty of the site may be justified.

*c. Topographic considerations.* Topography must be considered if maximum utilization of gravity flow through the entire system is to be achieved. Many recreational areas are well drained and gently sloping. Flat terrain usually requires a decision concerning pumping of wastewater to some point within the plant before adequate gravity flow can be obtained. Additional pumping costs may be necessary for a treatment facility on a site remote from visitor concentrations.

*d. Geologic and hydrogeologic considerations.*

(1) The capacity or incapacity of geological formations underlying the recreational facilities to support loads must be considered when selecting a site. Rock formations directly affect the excavation costs. The absorptive capacity of underlying soils is an important site selection parameter for various treatment systems. For example, land disposal systems require soils with high permeability for effective treatment. However, lagoons or other wastewater treatment processes that use earthen dikes should not be constructed over highly permeable soils, and they must be lined to avoid excessive rates of seepage from the basins. To avoid groundwater contamination, seepage rate should generally not exceed 0.3 mm/d ($^1/_8$ in/d).

(2) Adequate soil exploration is essential in site selection to guard against excessive seepage and against structural failure. Selected references are available to determine soil characteristics and expected properties (Taylor 1963, Teraghi 1960).

*e. Atmospheric condition considerations.* The atmospheric conditions of a candidate site must be evaluated during the planning phase; these include temperature, pressure, air movements, humidity, cloudiness, and precipitation. Average, as well as extreme, atmospheric conditions and variability of elements are also important considerations during site selection. Generally, it is best to locate recreational treatment facilities downwind from visitation centers to minimize odor and aerosol problems. If the construction of a recreational treatment facility at a remote site is not feasible, the design engineer must consider other alternatives, such as installing a landscape and/or decorative screen around the treatment plant and limiting the odor from the plant under normal operating conditions. Location is especially important where treated wastewater effluents are disposed by land application. For specific atmospheric condition considerations and requirements, see TM 5-814-3, Appendices D and E.

# Chapter 3
# Wastewater Generation And Characterization

## 3-1. General

This chapter provides generally available data that can be used to calculate water usage and wastewater generation, and to characterize the wastewater in terms of typical pollutant concentrations and characteristics.

## 3-2. Visitation and Length of Stay

*a. Capacity calculations.* Visitation (percent occupancy) and length of stay are important to consider when calculating the capacity of a recreational wastewater treatment system. Because visitation and length of stay are affected by factors such as season, climate, nearness to population centers, and types of facilities, the design engineer should base the capacity calculations on existing or projected visitation records, which are typically maintained by the recreation area manager.

*b. Direct calculations.* In the absence of such records, visitation data may be obtained by direct head count, admission fees, trailer count, and traffic count. Caution must be exercised when using traffic count because of internal movement of automobiles from area to area as well as outside traffic passing the check point. If outside traffic automobiles are included in the vehicle volume count, it will result in double counting the number of visitors.

## 3-3. Variations in Visitation

Visitation at recreational areas fluctuates vastly from season to season and from day to day within peak season. Percent occupancy should be used to calculate the maximum treatment system capacity. Percent occupancy can be estimated from historical records, where available, and by using equation (3-1). Table C-1 presents visitation data obtained from typical USACE recreational areas (Francingues 1976, Middleton USAEC). Where historical data are not available, equivalent population factors must be used as specified in TM 5-814-3, Chapter 4.

$$\text{Percent Occupancy} = \frac{\text{Recorded Income (\$/campsite/period)}}{(\text{No.Days/period}) \times \text{Campsite Fee (\$/day)} \times \text{No.Campsites} \times 100} \tag{3-1}$$

## 3-4. Water Usage and Wastewater Generation

*a. Overview.* The complexity of human activities in recreational areas makes estimating water usage and wastewater generation a difficult task. Table C-2 lists the facilities that typically exist at recreational areas which contribute to water usage and wastewater generation flows. The design engineer must account for the wastewater generated from all possible sources. Data for water usage and wastewater generation at typical USACE recreational areas are presented in Table C-3 (Metcalf & Eddy 1972). In addition, data for specific types of recreational area establishments including marinas are presented in Table C-4 (Corbitt 1990). Table C-5 lists comparative water use rates for various home appliances such as automatic dishwashers and garbage disposals (EPA-625-R-92/005, Matherly 1975, and Metcalf & Eddy 1972).

*b. Flow estimation methods.* There are two basic approaches used to estimate wastewater flows from recreational areas: the fixture unit method and the per capita method.

(1) Fixture unit method.

(a) Before using this method, the design engineer should obtain data on the number of fixture units at the site. Table C-6 lists the minimum number of sanitary fixture units required per site type (Penn Bureau of Resources, USDOI 1958). For marinas and other places where boats are moored, this number is based on the total number of seasonal slips and/or the number of transient slips, as appropriate. Sanitary facilities for marinas should be located conveniently within 152 m (500 ft) walking distance from the shore end of any dock. These sanitary facilities must be appropriately marked with signs readily identifiable.

(b) The data shown in Table C-7 can be used to estimate the wastewater flow based on the number of fixture units (Penn Bureau of Resources, USDOI 1958). (It should be noted that the data presented in Table C-7 represent hourly rates and are not directly related to fixture units as used in the plumbing codes to determine pipe sizes.) When using the fixture unit method, allowances should be made for special features such as trailer hookups, holding tanks, etc. Caution must also be used when applying the fixture unit method to estimate wastewater flows as this method is valid only when the number of fixtures is properly proportioned to user population. For user areas with minimum fixture comfort stations and a high percent occupancy, the fixture unit method may produce an underestimate of the wastewater flow.

(2) Per capita method.

(a) Table C-3 presents data which can be used to predict wastewater flows based on the per capita generation rate. The unit flows presented in Table C-3 are in agreement with water usage rates at various USACE recreational areas. The data presented in Table C-4 can be used as an additional design guide where site-specific flow data are not available. In computing wastewater flows from sanitary facilities servicing marinas only, assume for this method that each boat slip is equivalent to two persons.

(b) In addition, for marinas or other places where boats are moored which have a boat launching ramp and provide boat trailer parking space only while the boat is in use, the design flow must be increased by 38 L/d/capita (10 gal/d/capita) per boat trailer parking space. Where restaurants or motels are operated in conjunction with a marina or other place where boats are moored, the following will be used to determine the design wastewater flow:

- Motels: 246 L/d/capita (65 gal/d/capita) per constructed occupant space or a minimum of 492 L/d/room (130 gal/d/room).

- Restaurants: 190-680 L/d/customer seat (50-180 gal/d/customer seat). Each installation must be evaluated according to local conditions.

## 3-5. Monthly and Daily Flow Distribution

*a. Monthly flow distribution.* Monthly flow distribution at a specific site should be based on historical records or on flow data from a reasonably similar site. If these data are not available, then the general flow distribution shown in Table C-8 can be used. The monthly flow distribution data presented in Table C-8 are representative of recreational areas at inland reservoirs with moderate climatic conditions similar to those of the mid-Mississippi valley (Francingues 1976).

*b. Daily flow distribution.*

(1) The daily flow distribution is directly related to the percent occupancy on weekdays and weekend. The maximum daily flows can be estimated by both the fixture unit method and the per capita method.

(2) Weekend day (maximum). If using the fixture unit method, assume the maximum utilization of all fixtures and use the factors presented in Table C-7. For the per capita method, use predicted visitation data for the busiest month and the factors presented in Table C-3.

(3) Weekday (maximum). For both methods, assume 30-80 percent of the values obtained for weekend day. To select the appropriate value, consider the relative number of visitors on weekends compared to weekdays.

## 3-6. Wastewater Characterization

Wastewater from recreational areas can be characterized either as waterborne wastes such as those from picnic and camping areas, or as specialty wastes such as those from areas which use vaults, holding tanks, sanitary disposal (dump) stations, etc.

*a. Waterborne wastes.* Typical characteristics of waterborne recreational wastes are summarized in Table C-9 (Francingues 1976, Matherly 1975, Metcalf & Eddy 1972, and USAEWES). The concentrations of different pollutant parameters are not significantly different from those of domestic wastewater except for TKN and ammonia nitrogen ($NH_3$-N). It should be noted that wastewater characteristics may differ from facility to facility within a given recreation area. For example, picnic areas typically produce wastewater with higher nitrogen concentrations than do camping areas.

*b. Specialty wastes.* Identifying the sources and the characteristics of specialty wastes is an important element in the selection of the treatment process. Specialty wastes are generated from three sources: vaults, dump stations, and fish cleaning stations.

(1) Vault wastes.

(a) Vault wastes or septage from pit privies can be grouped into four categories: septic tank sludge (septage), vault waste, recirculating and portable chemical toilet waste, and low-volume flush waste. The organic strength, solids content, and chemical composition for these waste types must be known. Table C-10 presents the typical characteristics of a 3800-L (1000-gal) load of nonwater carriage waste (Smith 1973).

(b) Vault wastes with chemical or oil recirculating toilets are estimated to have the same organic characteristics as a standard vault (nonleaking), as reported by U.S. Forestry (Simmons 1972). Table C-11 summarizes the common pollutant parameters of vault wastes (Harrison 1972 and Simmons 1972). Vault wastes characterization data from other sources are summarized in Table C-12 (USAEWES). As can be seen in Tables C-11 and C-12, significant differences exist in the chemical (COD) and biological ($BOD_5$) composition of vault wastes. The $BOD_5$ and the COD concentrations in vault wastes depend upon detention time, dilution water entering the vault, and chemical additives.

(c) The values shown in Tables C-10 and C-11 are from primitive camping sites where a small amount of dilution water enters the vault with short detention times. These values may be considered as maximum composition values for vault wastes. The data in Tables C-11 and C-12 were obtained from areas receiving considerably more use, longer detention times, and large volumes of dilution water which accounts for the lower $BOD_5$ values. The addition of chemical oxidizing and liquefying agents contributes to lower $BOD_5$ values. Suggested design values for $BOD_5$ are: (1) the high values to be used for watertight vaults with no chemical additives; (2) 8800 mg/L for vaults without chemical additives which are frequently pumped and with moderate dilution; (3) 2500 mg/L only when chemical agents are added to the vault.

(2) Dump station wastes.

(a) Dump station wastes are basically generated from travel trailer and recreational watercraft wastes. Many travel trailer and recreation watercraft manufacturers have installed low-volume water flush and chemical recirculating toilets with holding tanks for trailer and boat wastes. Because indiscriminant dumping of these wastes into waterways, along highways, and at recreational areas is prohibited, the installation of sanitary disposal stations for boats and travel trailers is necessary.

(b) Availability of adequate treatment for sanitary wastes from boats and travel trailers is a major problem in most recreational areas (Robin and Green). Pump-out facilities are often many miles from the collection system of municipal treatment plants. The treatment of dump stations waste by conventional biological methods is not reliable because of the potential toxic effects of some chemical additives. Without large dilution, these wastes may cause treatment process upsets or otherwise affect treatment process efficiency. After a heavy weekend of recreational activity, shock loadings of dump station wastes have been shown to disrupt small municipal treatment plants (Robin and Green). Therefore, holding tanks, special treatment facilities, or arrangement for off-site treatment should be provided for dump station waste. Methods for special treatment include dilution of the biological and chemical load, equalization, and chemical treatment to neutralize toxic pollutants.

(c) The National Small Flow Clearinghouse (NSFC) has compiled a document outlining recent studies by researchers and scientists regarding the effects of chemical and biological additives on septic systems (NSFC-1). This document also lists additive manufacturers.

(d) The characteristics of the wastewater from 11 sanitary dump stations are summarized in Table C-13 (AOAC 1982, USEPA-1). A study of wastes from recreational water crafts revealed that these wastes also are highly concentrated, deeply colored, and contain variable amounts of toxic compounds (Robin and Green). The characteristics of typical waste pumpage from recreational water crafts are presented in Table C-14. Based on this study, it was concluded that arsenic, beryllium, molybdenum, or selenium were not detected in any of the 64 samples analyzed (Robin and Green). Mercury was detected in six samples at concentrations ranging from 6 to 9 mg/L. Relatively low concentrations (less than 0.2 mg/L) of cadmium, copper, manganese, nickel, and silver were found in most samples. Significantly high concentrations of aluminum, calcium, magnesium, tin, potassium, iron, and sodium were found. Toxic levels for certain metals were detected in individual samples as follows: cadmium as high as 104 mg/L, lead 79 mg/L, zinc 3540 mg/L, and copper 133 mg/L (Robin and Green).

(3) Fish cleaning station wastes. Typical characteristics of wastewater from fish cleaning stations are summarized in Table C-15 (Matherly 1975).

c. *Septage.* Septage is generally considered as the collection of sludge, scum, and liquid pumped from a septic tank. A broader definition might include any combination of liquid/solid waste retrieved from pit privies, vault, or other remote collecting or holding tanks. Septage generally contains hair, grit, rags, stringy material, and/or plastics and is highly odorous. Suspended solid concentrations in septage are as high as 5000 mg/L of inert material and 10 000 mg/L of volatile suspended matter.  Total solids have been reported at 15 000 mg/L of inert material and 25 000 mg/L of volatile solids (WEF MOP-8).

# Chapter 4
# Collection Systems

## 4-1. General

The purpose of a wastewater collection system is to convey wastes from the point of generation to the point of treatment or disposal. Depending on site conditions and economics, collected wastewater is conveyed either by truck transport or by piping system. The piping system may employ gravity, pressure, vacuum, or a combination of the first two. *Graywater* is defined as all wastewater produced from an occupied building unit (shower, bath, stationary stands, or lavatories) and generated by water-using fixtures and appliances, excluding the toilet and possibly garbage disposal, if any. *Blackwater* refers to pit privy waste and consists primarily of human excreta.

## 4-2. Absence of Pressurized Water Supply

When no pressurized water is available or soil conditions are unsuitable for direct ground disposal, the choice for onsite treatment may be limited to privies or waterless toilets. A privy, an outhouse over an earthen pit, is the simplest solution. When the pit is full, the privy may be closed or relocated. If the soil conditions are such that contamination of a groundwater source is a potential problem, impervious pits may be used and the subsequently collected waste (septage) pumped out and transported to a central holding tank or station. Both types of privies have been widely used for unserviced campgrounds, parks, and recreational areas without pressurized water service.

## 4-3. Transport by Truck

*a. General.* Trucks are used to transport four types of wastes: septic tank sludge, vault wastes, recirculating and portable chemical toilet wastes, and low-volume flush wastes. Factors to consider when designing a truck transport system include length of haul to the treatment facility, frequency of hauls, and the effect that the trucked waste has on the treatment facility (Clark 1971).

*b. Effects on treatment facility.*

(1) Table C-10 presents the characteristics of a 3800-L (1000-gal) load of nonwater carriage wastes. Addition of this waste type to a conventional treatment facility, without dilution, would adversely affect its efficient operation. Three parameters to be considered in developing dilution criteria for such wastes include solids concentration, presence of oxygen-demanding substances, and toxic chemical additives.

(2) Addition of truck-transported concentrated wastes to any treatment facility affects the equilibrium of a biological process. Operational procedures such as loading and wasting factors of the receiving wastewater treatment plant must be altered to accommodate the increase in solids concentration. To avoid an upset to the biological process equilibrium, the design engineer must estimate the amount of dilution required such that the sudden increase in mixed-liquor solids does not exceed 10 to 15 percent.

(3) Dilution and increased aeration capacity are both required to avoid the depletion of plant oxygenation capacity. Tradeoffs between dilution and increased aeration must be considered in order to treat concentrated wastes with minimal upset to the treatment system. For the waste types shown in Table C-10, the following dilution ratios may be used (USDHEW 1967):

- 19 parts water to 1 part septic tank waste

- 59 parts water to 1 part vault waste

- 44 parts water to 1 part low-volume waste

- 59 parts water to 1 part chemical toilet waste

c. *Design considerations.* To estimate the total amount of solids a system can tolerate, multiply the total amount of mixed-liquor volatile suspended solids by 10%. Table C-10 and the dilution factors shown above can be used to calculate the number of 3800-L (1000-gal) truckloads a conventional plant should receive. Therefore, frequency of truck transport can be estimated as a function of the receiving plant capacity.

d. *Operational considerations.* If a specifically designed wastewater treatment facility receives trucked wastes on a regular basis, oxygen demand becomes the limiting factor. If the treatment plant receives wastes on an irregular basis, both the solids equilibrium and the oxygenation capabilities must be considered. If the waste contains toxic chemical additives, maintaining the solids equilibrium should provide adequate dilution.

e. *Requirements for a transfer facility.* Primary requirements of a transfer facility include adequate storage capacity, ease of pumper truck unloading, comminution, odor control, and pumping flexibility and reliability. A typical truck unloading site contains a large discharge chute, bar screens, comminutors, and pressure water connections for flushing the truck after each dump. Transfer tanks should be equipped with dual pumps for reliability.

f. *Holding tanks or septage receiving stations.* Wastewaters from several pit privys, vaults, or small toilet systems may be temporarily held in a central holding tank or septage receiving station and then transported off-site for subsequent treatment and disposal.

(1) Considerations of design include adequate sizing with a liquid holding capacity of 7 to 14 days and a minimum capacity of 9500-L (2500 gallons); no discharges permitted from the tanks other than by pumping; a high-water alarm provided with allowances for a 3- to 4-day additional storage after activation; and the tank must be readily accessible to vehicles for frequent pumping. Since a holding tank constructed in or near fluctuating groundwater strata will be subject to flotation forces when the tank is evacuated or pumped clean, these considerations must be addressed in the holding tank's structural design.

(2) Septage receiving stations usually consist of an unloading area, reinforced-concrete septage storage tank and one or more grinder pumps, and a dry well on the effluent or pumping side of the septage wet well. Storage tanks are provided to store solid organic material to be disposed to an off-site treatment facility. The tank should be covered for odor control. If pretreatment (grit and screens) is not provided before storage, the tank should be equipped with influent grinder pumps to macerate any accumulated large solids. Chemical treatment (chlorine or lime) equipment can be provided if it is concluded in advance that the septage will require treatment, neutralization, or odor reduction.

(3) Design considerations for septage receiving stations include pressure hoses and washdown equipment; watertight truck hose connections and quick-release discharge tubes for the hose connections;

provisions for heater cable installations in the concrete chamber bottom to prevent freezing in colder climates; and a sloped ramp to tilt the pumper tank truck for complete discharge of contents.

(4) Pumping station designs should include a fail-safe arrangement for preventing pumper tank trucks from releasing septage without proper hose connections. In areas with a number of septic tanks or other individual sanitary facilities to be serviced, it is often difficult to discharge from pumper trucks unless a receiving station or holding tank is part of the overall septic collection system. Therefore, septic receiving and storage facilities with separate screening and grit removal constitute the best design arrangement. Generally, 100 mm (4 in) but preferably 150 mm (6 in) diameter lines are the minimum size for handling, receiving, and discharge lines.

(5) Design information for septage receiving stations can be found in Metcalf & Eddy 1991 and WEF MOP-8.

## 4-4. Gravity Flow Systems

*a. General.* Gravity flow systems consist of a network of underground sewer pipes sloping continually downhill to the wastewater treatment facility. Gravity systems must incorporate lift stations in order to avoid deep excavation that would be required in a flat or undulating terrain. It is desirable that piping systems be designed to avoid the formation of septic conditions, i.e., the velocity of wastewater through the piping system must be maintained to avoid the formation of septic conditions. The result of septic conditions is the formation of hydrogen sulfide, which causes odor and may cause damage to the piping materials. Therefore, maintaining a minimum flow of fresh wastewater is an important consideration when formulating a piping collection system.

*b. Design of gravity sewer systems.* Design information for gravity sewer systems can be found in Metcalf & Eddy 1991, TM 5-814-1, and WEF MOP-11.

*c. Manhole design.* Design information for manholes can be found in TM 5-814-1.

*d. Materials of construction.* Design information and guidance for the selection of materials for sanitary sewer construction can be found in TM 5-814-1.

*e. Installation and testing.* Design guidelines for sewer system layout and protection of water supplies can be found in TM 5-814-1.

## 4-5. Force Main Systems

*a. General.*

(1) Recreational areas may require pumping of wastewater from the point of generation to the point of treatment or disposal. Pumping is necessary when gravity flow is not practical due to topography and/or economic considerations, when there is insufficient head for gravity flow through a treatment system, or when the plant effluent must be lifted into the receiving stream or body of water. More details on general site selection requirements can be found in TM 5-814-2.

(2) There are two types of force main pressure systems: positive pressure and vacuum pressure. Table 4-1 presents a comparison of advantages and disadvantages of the two types of pressure systems.

*b. Location.* Guidance on location of pumping stations can be found in TM 5-814-2.

Table 4-1
Comparison of Pressure Systems

| Positive Pressure Systems | | Vacuum Pressure Systems | |
|---|---|---|---|
| Advantages | Disadvantages | Advantages | Disadvantages |
| Eliminate the need to lay pipe to hydraulic grade lines. | Wastewater pumps are required at every sewage input point to lift the wastewater into the network of collection lines. | Use a central pumping station to maintain vacuum on the main line. | Length of pumping possible due to head limitations. |
| Eliminate the need for lift stations of a conventional system. | Require electrically operated mechanical equipment at every sewage input point. | Require a normally closed (NC) valve at each point of sewage input. | |
| Substitute the small-diameter plastic pipe for large diameter pipe. | Solids must be broken up by providing either a grinder pump or other comminution units at each entry point. | Collection lines are small-diameter pipes that can be laid without regard to hydraulic grade lines. | Possible vacuum leaks that render the system inoperable. |
| Infiltration is eliminated because manholes are not required, thus piping materials are not exposed to groundwater fluctuations. | | Reduction in quantity of flushing water needed. | |

c. *Materials of construction.* Design information and guidance for the selection of materials for force main pressure sanitary sewer construction can be found in TM 5-814-1 and TM 5-814-2.

d. *Installation and testing.* Guidelines for force main pressure sanitary sewer system site selection and building and site requirements can be found in TM 5-814-1 and TM 5-814-2.

e. *Pumping equipment.* Four basic types of pumps are employed in wastewater collection systems: centrifugal pumps, screw pumps, pneumatic ejector pumps, and grinder pumps. Descriptions and general design specifications for each pump type can be found in TM 5-814-2.

f. *Pump selection.* Design guidance for pumping systems design and pump selection can be found in TM 5-814-2.

g. *Wet well requirements.* Guidance for wetwell design can be found in TM 5-814-2.

h. *Pump station components.* Guidance for pump stations construction and components design can be found in TM 5-814-2.

## 4-6. Alternative Wastewater Collection Systems

a. *System types.* As the cost of conventional gravity sewer collection systems sometimes exceeds the cost of wastewater treatment and disposal facilities, it has become necessary to develop alternative sewer collection systems. Current alternatives to conventional gravity collection systems include positive pressure sewer systems, vacuum sewer systems, and small-diameter gravity sewers. Alternative sewer collection systems are applicable to remote or recreational areas. However, the final selection of an alternative wastewater collection system should be based on economic considerations.

b. *Examples.* Design examples of the three alternative wastewater collection systems can be found in EPA/625/1-91/024.

# Chapter 5
# Treatment Design Considerations

## 5-1. General

This chapter identifies treatment design considerations for wastewater treatment facilities and/or plants with relatively small capacities, those constructed in recreational areas with wastewater flows less than 379 000 L/d (100,000 gal/d). (Note: these considerations generally apply to large treatment facility planning as well.) Factors to be considered in the preparation of a design for small wastewater facilities include site selection, treatment system selection, and design steps. Site selection considerations are presented in Chapter 2, and certain design steps and process selection criteria are to be found throughout Chapter 7. For proper design it is mandatory to know the quantity of wastewater to be expected (see paragraph 3-4), the monthly and daily flow distributions (see paragraph 3-5), as well as the wastewater composition, constituents, and strength (see paragraph 3-6).

## 5-2. Small Individual Units

*a. Pit privy.*

(1) Historically the pit privy has been the simplest and most commonly used wastewater treatment device. It is a non-water carrying unit which has been developed to store human waste from a single building or several small buildings without other sanitary facilities. In brief, a pit privy is a dug hole over which an outhouse has been built. Privy construction should be limited to low use or highly remote areas, as for all intents and purposes, privies have been replaced by "Port-a-Johns" or chemical toilets which may be easily transported by truck from one location to another during periods of high use. To be effective, pit privys must be pumped out from time to time and the septage trucked to a larger holding tank or wastewater treatment plant (see paragraph 4-3). The privy waste, consisting primarily of human excreta, is generally referred to as "blackwater," as opposed to the "graywater" generated by water-using fixtures such as showers and lavatories.

(2) Design considerations for pit privies include additional requirements such as animal- or rodent-proofing. Privy contents should not be permitted to overflow onto the ground surface, and surface drainage should be directed away from the privy site. The privy site should be constructed on raised concrete slab, and preferably located well above the underlying groundwater table. The privy structure should be constructed of durable wood or molded plastic and built to last 10-15 years.

(3) There is no generally accepted standard privy design. Unlined pits of short length, width and depths are simply dug and covered with a fabricated (plastic or wooden) structure resting on a concrete slab with apertures or holes in wooden or plastic seats. For general information regarding options and guidelines for pit privies consult USDA-1 and USDA-2.

*b. Vault toilets.*

(1) Simple vault toilets, or outhouses over enclosed chambers, are most often used for remote-site wastewater treatment. The toilets are located in vented structures under which is a below-ground enclosed and preferably watertight chamber fabricated to prevent both infiltration and exfiltration. The terms vault toilet and pit privy are often used interchangeably: both must be periodically pumped out; both have associated odor problems; both are the receptacles for rags, cans, trash, bottles, plastic, meal containers,

and almost any throw-away objects or items carried by users; both attract disease vectors; both have the potential for contaminating the groundwater resources; both require frequent oversight; both often require addition of chemical additives. Except under unusual circumstances, use of pit privys and vault toilets should be discouraged when construction of modern wastewater treatment facilities is being considered.

(2) Improved aerated vault toilets have been in operation at a number of Army facilities for over 20 years. Various types of air compressors and blowers, including diffuser types, have been successfully used at these facilities. The two most accepted types of aeration system configuration are bubble aeration and mechanical aeration. Bubble aerators are belt-driven, lubrication-free, carbon-vaned blowers. Blower inlets must be provided and fitted with a replaceable-element air filter. Blower outlets must be connected to a perforated air distribution pipe mounted along the vault floor. Air must be continuously supplied to mix wastes and supply oxygen.

(3) The alternative method for aerating a vault toilet is mechanical, i.e., mixing of the wastes using a motor-driven impeller combined with injection of air below the vault liquid surface. The entire unit is mounted on a float which rides on the waste surface, thereby maintaining a constant immersion depth for the impeller. When operating, the impeller creates a vortex which lowers the pressure at the end of the hollow shaft driving the impeller and allows the atmospheric pressure to draw air down the shaft and into the liquid waste where it is mixed by the vortex. Of the two systems, the bubble aeration system appears to have fewer design and operating problems.

(4) Both bubble and mechanical type aeration devices require electric power. Unlike composting toilets (see paragraph 5-2c below), these two systems require too much energy on a continuous basis to make solar power practical.

(5) General design information for vault toilets can be found in USACERL 1984, USDA-1, and USDA-4. General information regarding options and guidelines for the selection of vault toilets can be found in USDA-2 and USDA-4.

   c.  *Composting toilets.*

(1) Composting is the controlled decomposition of organic material into humus. The organic materials are converted to a more stable form by either aerobic decomposition or anaerobic fermentation. Most composting toilets are designed for continuous aerobic decomposition of human waste. As flushing of waste is not provided for, no water is introduced into the composting chamber, which receives fecal matter, urine, toilet tissue and a bulking agent (sometimes sawdust). Composting generally decreases the volume of the waste. Electricity must be made available for ventilation. Ventilation consists of a vent pipe and fan system to remove carbon dioxide, water vapor, and air from the composting chamber. Composting toilets have capacities for 2 to 25 persons with 2- to 6-person capacities being the most common. Electrical heating elements are usually provided for cold-weather climates. Where solar energy is available or feasible, it should be used for heating and ventilation (Clivus Multrum, a patented process). Composting toilets may be used as alternatives to pit privies, vault toilets, or chemical toilets.

(2) General design information for composting toilets can be found in USACERL 1984, USDA-4, and USDA-5.

*d. Septic tank systems.*

(1) The septic tank has been successfully employed for well over a century, and it is the most widely used on-site wastewater treatment option. Septic tanks are buried, watertight receptacles designed and constructed to receive wastewater from the structure to be served. The tank separates solids from the liquid, provides limited digestion of organic matter, stores solids, and allows the clarified liquid to discharge for further treatment and disposal. Settleable solids and partially decomposed sludges accumulate at the bottom of the tank. A scum of lightweight material (including fats and greases) rises to the top of the tank's liquid level. The partially clarified liquid is allowed to flow through an outlet opening positioned below the floating scum layer. Proper use of baffles, tees, and elbows protects against scum outflow. Clarified liquid can be disposed of to soil absorption field systems, soil mounds, lagoons, or other disposal systems.

(2) Factors to be considered in the design of a septic tank include tank geometry, hydraulic loading, inlet and outlet configurations, number of compartments, temperature, and operation and maintenance practices. If a septic tank is hydraulically overloaded, retention time may become too short and solids may not settle properly.

(3) Both single-compartment and multi-compartment septic tank designs are acceptable. Baffled or multi-compartment tanks generally perform better than single-compartment tanks of the same total capacity, as they provide better protection against solids carryover into discharge pipes during periods of surges or upsets due to rapid sludge digestion. Poorly designed or placed baffles create turbulence in the tank which impairs the settling efficiency and may promote scum or sludge entry into the discharge pipes.

(4) Septic tanks, with appropriate effluent disposal systems, are acceptable where permitted by regulatory authority and when alternative treatment is not practical. When soil and drainage characteristics are well documented for a particular site, septic tank treatment is eminently feasible for small populations. Septic tanks are effective in treating from one to several hundred population equivalents of waste, but should generally be used only for 1 to 25 population equivalents, except when septic tanks are the most economical solution for larger populations within the above range. The minimum tank size is at least 1 900-L (500-gal) liquid capacity. In designing tanks, the length-to-width range should be between 2:1 and 3:1, and the liquid depth should be between 1.2 and 1.8 m (4 and 6 feet). When effluent is disposed of in subsurface absorption fields or leaching pits, a minimum detention time in the tank based on average flows is generally required. Different states have specific detention time requirements. Table B-2 identifies the states which have specific septic tank design requirements.

(5) The septic tank must be sized to provide the required detention (below the operating liquid level) for the design daily flow plus an additional 25 percent capacity for sludge storage. If secondary treatment, such as a subsurface sand filter or oxidation pond or constructed wetland is provided, the detention period may be reduced. Open sand filter treatment of septic tank effluent can further reduce the required detention time. Absorption field and leaching well disposal should normally be limited to small facilities (less than 50 population equivalents). If the total population equivalent is over 50, then more than one entirely separate absorption field would be acceptable. For ten or more population equivalents, discharge of effluent will be through dosing tanks which periodically discharge effluent quantities of up to 80 percent of the absorption system capacity.

(6) Combined septic tank and recirculating sand filtration systems have been shown to be effective in providing a closed-loop treatment option for either a single or a large septic tank or a multiple series of small tanks. The septic effluent is directed to a recirculation tank from which it is discharged by a sump

pump to a sand filter contained in a open concrete box which permits no seepage to the local groundwater table. The recirculation pump discharge enters a multiple-pipe distribution device of manifold-lateral design overlying the sand bed. The discharge is sprayed, or trickles, onto a sand bed underlaid by graded gravel through which underdrain piping collects the filtered wastewater and returns to the same recirculation tank. Ultimately, the wastewater can be discharged to a receiving stream following disinfection, or to a leaching or absorption field. The design of a septic tank system discharging to surface waters normally includes sand filtration to lower the concentrations of suspended solids and $BOD_5$ to acceptable levels. Ammonia-nitrogen concentration in the effluents may be high, especially during winter months.

(7) Design features for various septic tank systems including leaching or adsorption fields and mound systems can be found in Burs 1994, Converse 1990, EPA/825/1-80/012, EPA/625/R-92/010, Kaplan 1989, Kaplan 1991, OSU 1992, and Perkins 1989.

*e. Imhoff tanks.* The Imhoff tank is a primary sedimentation process which performs two functions, the removal of settleable solids and the anaerobic digestion of those solids. In these respects, the Imhoff tank is similar to a septic tank. The difference is that the Imhoff tank consists of a two-story tank in which sedimentation occurs in the upper compartment and the settled solids are deposited in the lower compartment. Solids pass through a horizontal slot at the bottom of sloping sides of the sedimentation tanks to the unheated lower compartment for digestion. Scum often accumulates in the sedimentation chamber, where it may be skimmed off manually. Digestion-produced gases rise vertically, and with a properly designed overlapping sloping wall arrangement, are directed through length-wise vents on either side of the horizontal sedimentation chamber. Thus, gases and any entrapped sludge particles rise to the upper compartment liquid surface from the bottom sludge and do not interfere with the sedimentation process in the upper compartment. Accumulated or digested sludges are withdrawn from the lower compartment by hydrostatic head through a simple vertical piping system. Sludges must be disposed of after being dried on sand drying beds or other approved systems.

(2) Design features for Imhoff tanks can be found in Middleton USACE, WEF MOP-8, and Metcalf & Eddy 1991.

## 5-3. Conventional Wastewater Treatment Facilities

Conventional wastewater treatment refers to a complete biological wastewater treatment process that includes flow measurement and equalization, primary and secondary sedimentation treatment, biological treatment, and effluent disinfection. The following paragraphs present a discussion of the various unit processes and design considerations for conventional wastewater treatment processes as well as pretreatment considerations. A description of flow monitoring devices is presented in Chapter 6.

*a. Oil and grease interceptors.*

(1) When restaurants, laundromats and/or service stations are located within the sewer collection system, but away from the camping or tenting or recreational areas, the liquid wastes discharged to a treatment facility typically contain oil and grease, cleaning agents, and organic material from kitchen sink garbage disposal units which interfere with the treatment process effectiveness. Grease is usually collected in interceptor traps using cooling and/or flotation, while oils are intercepted by flotation. For effective flotation, a grease interceptor trap or and oil/water separator should be designed with a minimum detention time of 30 minutes (Metcalf & Eddy 1991).

(2) In remote areas where many small treatment plants have been constructed, oil- and grease-related operations and maintenance problems sometimes arise once the facilities are in full use. Therefore, the selection of effective low-maintenance solutions for collecting oils and greases is of paramount importance when designing a wastewater treatment system for such locations.

*b. Preliminary treatment.* Preliminary treatment is the conditioning of a waste stream to partially reduce or remove constituents that could otherwise adversely affect the downstream treatment processes. Preliminary treatment processes include coarse screening, comminutor, grit removal, and flow equalization.

(1) Coarse screening.

(a) Coarse screening includes both manually and mechanically cleaned bar racks. Mechanical screens have generally replaced the hand-cleaned racks. Manual labor is required to reduce rack clogging and clean the bar racks by vertically pulling the collected debris with rakes or tongs onto a perforated plate on top of the rack and then disposing of the rakings. Mechanically cleaned racks are divided into four types: chain operated; reciprocating rake; catenary; and cable. Some mechanically cleaned racks are cleaned from the upstream face and some from the downstream face. Chain, rake, catenary, or cable cleaning devices usually operate on set timing sequences and are not necessarily flow dependent.

(b) Screenings are the floating or suspended material collected and retained on bar racks. The quantity of screening retained increases with smaller openings between bars. Coarse screenings typically retained on racks or bars with spacings greater than or equal to 13 mm (0.5 in) include plastics, rags or fibrous materials, rocks, floating wood, lawn waste or plant cuttings, and other miscellaneous materials which find their way into sanitary sewers.

(c) Typical design information for hand and manually cleaned racks can be found in Droste 1997, Metcalf & Eddy 1991, Reynolds 1995, and WEF MOP-8.

(2) Comminutors.

(a) Comminutors are adjuncts to bar racks or screens and sometimes are alternatives to the coarse screening devices. Most contemporary designs consist of vertical revolving-drum screens. All designs, irrespective of efficiency, are equipped with high-quality metal cutting disks or teeth which periodically require sharpening.

(b) Comminutors are continuously operating devices for catching and shredding heavy, solid, and fibrous matter; the suspended material is cut into smaller, more uniform sizes before it enters the pumps or other unit processes. Some fibrous material which is shredded or cut may later recombine into ropelike pieces following comminution.

(c) Typical design information for comminutors can be found in Metcalf & Eddy 1991, Reynolds 1995, and WEF MOP-8.

(3) Grit removal.

(a) Grit is the heavy suspended mineral matter present in wastewater (sand, gravel, rocks, cinders), which is usually removed in a rectangular horizontal-flow detention chamber or in the enlargement of a sewer channel. The chamber may be aerated to assist in keeping the influent wastewaters from becoming septic. Detention reduces the velocity of the influent and permits separation of the heavier material by differential settling. A mechanical grit collection system may be provided to collect the grit and convey it to a point of collection adjacent to the grit chamber, usually a metal can or dumpster. Vortex-type grit chambers may also be employed. Manually cleaned, gravity grit chambers are not considered state-of-the-art and generally should not be included in modern wastewater treatment plant design.

(b) Typical design information for horizontal flow and aerated grit chambers can be found in WEF MOP-8, and in Metcalf & Eddy 1991 for vortex-type grit chambers.

(4) Flow equalization.

(a) Flow equalization is a method of retaining wastes in a separately constructed basin such that the basin effluent is reasonably uniform in flow and wastewater characteristics or strength. The purpose of equalization is to average, dampen, or attenuate the flow and composition of the waste stream. In effect, the equalizing or holding basin is a balancing reservoir. Equalization tanks may be either in-line, or single basin, in which the influent flows directly into and is directly drawn off from the basin; or sideline, which employs two basins, the larger usually serving as the prime equalization basin and the other as a pump wet well.

(b) Equalization basins may be placed off-line in the collection system, after headworks, or at a point to following primary clarification (and before advanced treatment, if any).

(c) Techniques for sizing equalization basins include the mass-diagram for hydraulic purposes, and statistical techniques and interactive procedures for hydraulic and organic conditions. Although mixing of basin influent is not always provided, mixing methods include surface aeration, diffused air aeration, turbine mixing, and inlet flow distribution and baffling.

(d) Typical design information for equalization can be found in Metcalf & Eddy 1991 and WEF MOP-8.

*c. Primary sedimentation treatment.*

(1) In a conventional wastewater treatment plant, primary sedimentation is employed to remove settleable particulate and colloidal solid material from raw wastewater. The principal design considerations for primary clarification or sedimentation basins are horizontal flow cross-sectional areas, detention time, side water depth and overflow rate. Efficiency of the clarification process is significantly affected by the wastewater characteristics, suspended solids concentration, the number and arrangements of basins, and variations in the inflow.

(2) Settling basin designs must provide for effective removal of suspended solids from the wastewaters which have already passed through the preliminary process units (screens, grit chambers, comminutors, equalization basins), and collection and removal of settled solids (sludge) from the basin. Short-circuiting of flows in sedimentation should be avoided whenever possible.

(3) Basin design should consider the following factors: basin inlet and outlet velocities; turbulent flow; wind stresses, if any; temperature gradients; and basin geometry.

(4) Principal design considerations should also ensure evenly distributed inlet flow with minimal inlet velocities to avoid turbulence and short-circuiting; quiescent conditions for effective particle and

suspension settling; sufficient basin depth for sludge storage to permit sufficient or desired thickening; mechanical sludge scrapers (horizontal or circular) to collect and remove the sludges; and minimum effluent velocities by limiting weir loadings and by proper weir leveling and placement. Plain sedimentation horizontal flow basins may be either circular or rectangular. The preferred minimum diameter of circular clarifiers in small conventional wastewater treatment plants is 3 m (10 ft), and a like dimension for the sidewater depth.

(5) Typical design information for primary sedimentation basins can be found in Droste 1997, Metcalf & Eddy 1991, Reynolds 1995, and WEF MOP-8.

    *d.  Secondary sedimentation treatment.*

(1) Secondary sedimentation or final clarification is employed to remove the mixed-liquor suspended solids (MLSS) following the activated sludge processes and oxidation ditches treatment or to remove growths that may slough off from trickling filters and rotating biological contactors. Well-designed secondary clarification processes produce high-quality effluents with low suspended solids. In advanced or tertiary treatment plants (rarely found in remote or recreational areas), secondary sedimentation is employed to remove flocculated suspended solids and/or chemical precipitates. The same design considerations for primary sedimentation apply to secondary clarification. Efficient sludge collection and removal from secondary sedimentation is of prime importance in the design procedure.

(2) Typical design information that applies to secondary sedimentation treatment can be found in Droste 1997, Metcalf & Eddy 1991, Reynolds 1995, and WEF MOP-8.

    *e.  Trickling filters.*

(1) The conventional secondary treatment trickling filter process employs an attached-growth biological system based on passing (trickling) organically loaded wastewater over the surface of a bio-logical growth attached to a solid media which is firmly supported on a well-ventilated underdrain system. The conventional trickling filter process is best employed in situations where the organic concentrations in the effluent from the primary sedimentation process are moderate rather than high.

(2) Trickling filters are generally classified, with respect to the application rate of both organic and hydraulic loadings, as low rate, high rate, and roughing or super rate. Super-rate or roughing filters are not applicable to wastewater plants at recreational areas and require special Corps of Engineers approval prior to construction. The process is further categorized by media type, media depth, number of trickling filter stages, mode of wastewater distribution (fixed nozzles in smaller units or rotary arm distributors), and/or intermittent dosing cycles or frequency.

(3) Recirculation of trickling filter effluent back through the primary sedimentation basins or directly to the trickling filter influent, or to the second filter in a two-stage system, is often practiced. The main purpose of recirculation is to provide continuous flow through the filter media to maintain a continuous organic material feed for the media-attached microorganisms and to prevent dehydration of the attached growth.

(4) The most frequently employed trickling filter media is granite rock. Slag has also been used. Plastic media in various shapes and redwood lath media are also to be found in more recently designed processes. The principal criteria for media are the specific surface area (area per unit volume) and the percent void space. Organic loading is directly related to specific surface area available for the

media-attached biological growth. Increased void spaces enhance oxygen transfer to the attached growths, provide adequate ventilation throughout the media bed, and permit significantly higher hydraulic loadings. Should a dosing system be required (infrequently installed in new designs and limited to low-rate filters), dosing of wastewater to the media should occur at least every 5 minutes to ensure or provide nearly continuous liquid applications. Ventilation of the media bed is necessary to ensure effective operation of the trickling filter. In both cold and warm climates, disinfection of trickling filter plant effluent is required.

(5) Design information for trickling filter systems, media, and rates of application can be found in Metcalf & Eddy 1991, Reynolds 1995, and WEF MOP-8.

*f. Extended-aeration activated sludge processes (package plants).* The activated sludge process, in conventional or modified forms, has been shown to meet secondary treatment plant effluent limits. The three modified categories of activated sludge processes for small wastewater plants are extended-aeration package plants, oxidation ditches, and sequencing batch reactors. All are primarily based on the food-to-microorganism (Food:Mass, or F:M) ratio principle.

(1) Extended-aeration package plant.

(a) The extended-aeration activated sludge process is commonly used to treat small wastewater flows up to 379 000 L/d (100,000 gal/d). The aeration process ranges in detention time from 24 to 36 hours and is, along with oxidation ditches (see below), considered the longest of any activated-sludge process. As $BOD_5$ loadings are generally low in recreational areas, the extended aeration system usually operates in the endogenous growth phase. Extended aeration processes generally accept periodic or intermittent heavy organic loadings without significant plant upsets.

(b) Process stability is believed to result from the large aeration tank volume as well as the continuous and complete mixing of tank contents. Long detention times and low overflow rates are effected by final settling tank design. Overflow rates generally range from about 8150 to 24 450 L/d/m$^2$ (200 to 600 gal/d/ft$^2$). Aeration tank volumes ranging from 19 000 to 379 000 liters (5000 to 100,000 gallons) are not uncommon, although this is considered unusually large for such small flows. Generally, excess sludge is not removed continuously because the suspended solids concentration of the mixed liquor is permitted to increase with intermittently periodic dumping of the aeration tank.

(c) The popularity of small extended-aeration plants has increased the demand for factory-manufactured and -assembled units. Most extended-aeration systems are factory built with the mechanical aeration equipment often installed in cast-in-place reinforced-concrete tanks.

(d) Components of extended-aeration package plants include a combination built-in bar screen and comminutor aeration basin or compartment, aeration equipment, air diffuser drop assemblies, a froth spray system, at least two air blower and motor combinations (rotary positive blowers are usually preferred), and a hoppered clarifier to receive aerated mixed liquor from the aeration compartment. Airlift educator pipes return hopper-collected sludge to a front-of-system sludge holding tank through a relatively small sludge waste line. A totalizer meter with a specific flume or weir is required for flow measurement.

(e) Seasonal changes are important in extended-aeration package plants as the efficiency of treatment will be determined, to a great extent, by the ambient temperature. During winter months in colder climates the activity of microorganisms is reduced, particularly in above-ground tanks. For example, if the organic loading remains constant, more microorganisms are needed in winter months than in summer months to achieve the same effluent quality. Lower temperatures also affect the dissolved oxygen concentrations in the aeration chamber; the colder the liquid, the more oxygen the wastewater can assimilate in solution. As the temperature increases, the ability of the wastewater to assimilate gases in solution decreases. Sludge production also varies with seasonal changes as the biological life forms are more active in warmer climates. In both cold and warm climates, disinfection of the plant effluent is required.

(f) Design information for extended-aeration package plants can be found in Metcalf & Eddy 1991, Reynolds 1995, and WEF MOP-8.

(2) Oxidation ditches.

(a) Oxidation ditches, also referred to in the U.S. Army as Closed Loop Reactors (CLRs), are activated sludge treatment processes of reinforced-concrete or steel tank design and are principally considered a secondary treatment process. Small prefabricated units of metal tank construction are commercially available. Depending on the anticipated wastewater composition, some of the preliminary units, particularly primary sedimentation, may be omitted. Oxidation ditches are extended aeration processes that also operate on a food-to-microorganism ratio principle. Reaction time, based on influent flow, varies from 18 to 36 hours, but typically averages 24 hours. The reactors are usually circular or "race-track" shaped. Brush-type aerators (rotating axles with radiating steel bristles) or vertically mounted shaft propeller aerators aerate the wastewater and provide the constant motion of the wastewater in the reactor. The vertical shaft aerators are designed to induce either updraft or downdraft. Final clarification or secondary settling is a required feature with an additional capacity to recycle activated sludge from the clarifier bottom. In both cold and warm climates, disinfection of the plant effluent is required.

(b) Design information for CLRs can be found in Metcalf & Eddy 1991, Reynolds 1995, TM 5-814-3, and WEF MOP-8.

(3) Sequencing batch reactors (SBRs).

(a) SBR systems combine biological treatment and sedimentation in a single basin. The design considerations for SBRs include the same factors commonly used for a flow-through activated sludge system. The principal operating stages of an SBR system include:

- Static fill—influent flow is introduced to an idle basin.

- Mixed fill—influent flow continues and mixing by diffused aeration begins.

- React fill—influent flow continues, mixing continues, and mechanical aeration begins.

- React—influent flow is stopped and mixing and aeration continue.

- Settle—mixing and aeration are stopped and clarification begins.

- Decant—clear supernatant is decanted.

- Waste sludge—optional sludge wasting may occur.

- Idle—basin is on standby to restart the process.

(6) SBRs also operate on the Food:Mass (F:M) ratio, which ranges generally from 0.05 to 0.30, and from 0.10 to 0.15 for domestic waste. At the end of decant the stage, the mixed-liquor suspended solids (MLSS) concentration may vary between 2000 and 5000 mg/L. A typical value for municipal waste would be 3500 mg/L. At least two basins are provided in an SBR design to provide operational flexibility and improve effluent quality. Design criteria information for SBRs can be found in EPA-625-R-92/005 and EPA/625/R-92/010.

g. *Rotating biological contactors (RBC).*

(a) The RBC is an attached growth secondary treatment process. RBC is principally composed of a box-like container, most frequently a concrete tank or metal vat, through which wastewater flows following preliminary treatment, and a complex of multiple plastic discs mounted on a horizontal shaft. The shaft is mounted at right angles to the wastewater flow; approximately 40 percent of the total disc area is submerged. As the shaft rotates, the disc slowly revolves and biological growths flourish on the disc plates by sorbing organic materials; these growths slough continuously off, thereby eliminating any excess growth. As the top 60 percent of the disc plate area passes through the air, oxygen is absorbed to keep the growths in a semiaerobic state. In principle, the mobile sorption and oxidation processes simulate the static trickling filter media growth conditions.

(b) Multistage RBC units consist of two or more stages in series. Multistage contactors achieve greater $BOD_5$ removal than do single-stage contactors. Recycling of RBC effluent to the head of the plant is usually not practiced. Any sludge collected in the holding vat, container, or tank, along with secondary clarifier sludge, is usually returned to the primary clarifier influent stream to aid in thickening of any collected primary sludge.

(c) In cold climates, discs and operating equipment are generally covered to reduce heat loss and to protect the system from freezing. In warmer climates, no permanent enclosed structure is required, although open-sided sunroofs may be provided.

(d) The principle RBC design consideration is the wastewater flow rate per unit surface of the discs. A properly designed hydraulic loading rate produces an optimum food-to-microorganism ratio. Peripheral speed of the discs is usually limited to 0.3 m/s (1 ft/s). Disinfection of the plant effluent is required.

(e) Design information for RBCs can be found in Metcalf & Eddy, Reynolds 1995, and WEF MOP-8.

## 5-4. Stabilization Ponds

a. *Classifications.* Stabilization ponds provide treatment for wastewaters through a combination of sedimentation and biological treatment using extensive detention times. Stabilization ponds are generally categorized as aerobic, facultative, or anaerobic according to their dissolved oxygen depth profile. An aerobic pond has varying concentrations of oxygen throughout its entire depth, while an anaerobic pond is devoid of oxygen at any depth except in the very top few millimeters (inches) at the air-liquid interface. A facultative pond supports oxygen, or is aerobic, in its top zone and is anaerobic or devoid of oxygen at the lower depths or bottom zone. Most stabilization ponds fall into the facultative category. The amount of oxygen present depends on temperature, organic loading, and sunlight intensity (photosynthetic effect), and the dissolved oxygen concentrations in the top zone will vary with time and conditions. During periods of no direct sunlight (prolonged hazy or very cloudy conditions) and during the night hours, dissolved oxygen concentrations will decrease as the dense microbe population and algae, if any, readily consume available oxygen. As the microbes and algae expire, they settle to the pond bottom and enter the anaerobic state and decompose. Few local odors are experienced in facultative and aerobic ponds; anaerobic ponds produce pronounced odors.

(b) Design guidance. Design considerations and guidelines for stabilization ponds and aerated lagoons can be found in Metcalf & Eddy 1991, Reynolds 1995, and WEF MOP-8.

## 5-5. Natural Systems for Wastewater Treatment

   *a.   System types and parameters.*

(1) Natural systems for land wastewater treatment encompass both soil-based and aquatic methods. These systems consume less energy and produce less sludge than conventional systems. Soil-based methods include subsurface systems such as septic tank leach fields to serve occupants of a single structure, limited populations in small communities, or a few visitors to remote areas. Aquatic methods are those in which wastewater is applied at the surface of the soil and include slow-rate (SR) land treatment, rapid infiltration (RI) land treatment; and overland flow (OF) land treatment. SR and RI systems rely on infiltration and percolation in soil matrices for applied wastewater movement. OF systems utilize sheet flow of the applied wastewater along a gentle slope. Vegetation is important in both SR and OF systems. RI hydraulic loading rates are higher than those for SR and OF systems, and vegetation is not important. Surface application gives the SR and RI systems a higher treatment potential than OF; most aerobic microbic activity occurs in the top layer of soil and not merely along the soil surface treated by the OF system. Figure 5-1 presents a process selection chart for natural systems (MOP FD-16, Natural Systems for Wastewater Treatment). Figure 5-2 presents a decision diagram for selecting wetland alternatives (MOP FD-16). Figure 5-3 identifies climatic control and zone considerations for land treatment facility selection (MOP FD-16).

(2) Limiting design parameters for the various natural system types are as follows:

- For on-site septic leach fields—hydraulic capacity of soil.

- For slow-rate land treatment—hydraulic capacity, nitrogen or phosphorous.

- For rapid infiltration land treatment—hydraulic capacity, nitrogen or phosphorous.

- For overland flow land treatment—$BOD_5$, TSS removal, infrequently nitrogen.

(3) For flows of 378 500 L/d (100,000 gal/d), SR processes may require 1 to 10 hectares (2.5 to 25 acres), OF processes from 0.4 to 2 hectares (1 to 5 acres), and RI basins from 0.4 to 1.2 hectares (1 to 3 acres) of suitable land surface.

   *b.   Slow rate (SR) land treatment.*

(1) SR is a widely used treatment method, and requires the highest level of pretreatment. Secondary clarification and disinfection are not uncommon prior to using SR. SR systems typically achieve the highest level of performance of the three natural systems. Site requirements depend on loading rates, site characteristics, and design objectives. Loading rates generally vary from 1 to 2 m (3 to 7 ft) of applied wastewater per annum, much of which can be lost by evapotranspiration. Additional area is required for any needed pretreatment systems, roads, odor buffer zones, and structures. Soil type and depth to

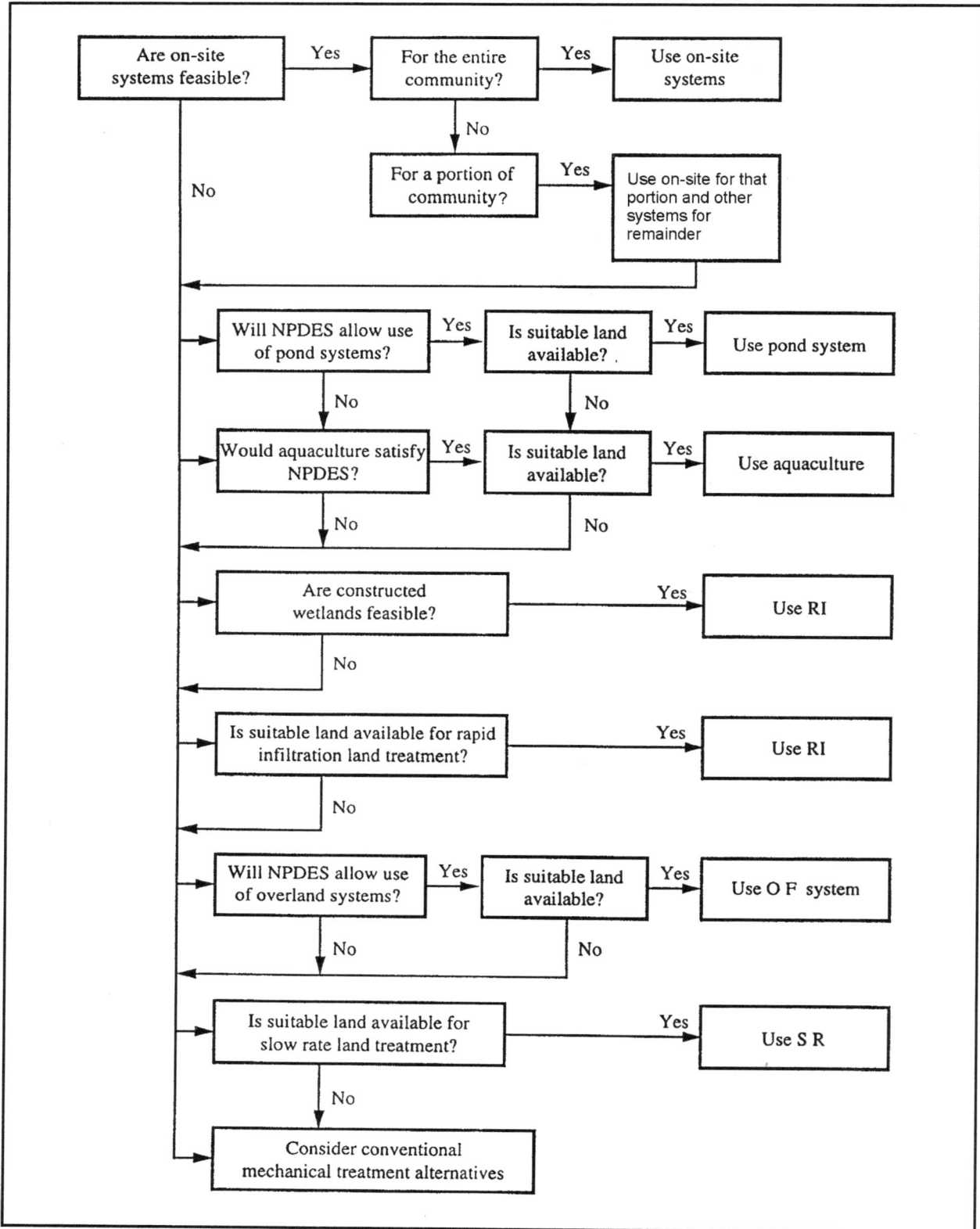

**Figure 5-1. Process selection for natural treatment systems (copyright © Water Environment Federation, used with permission)**

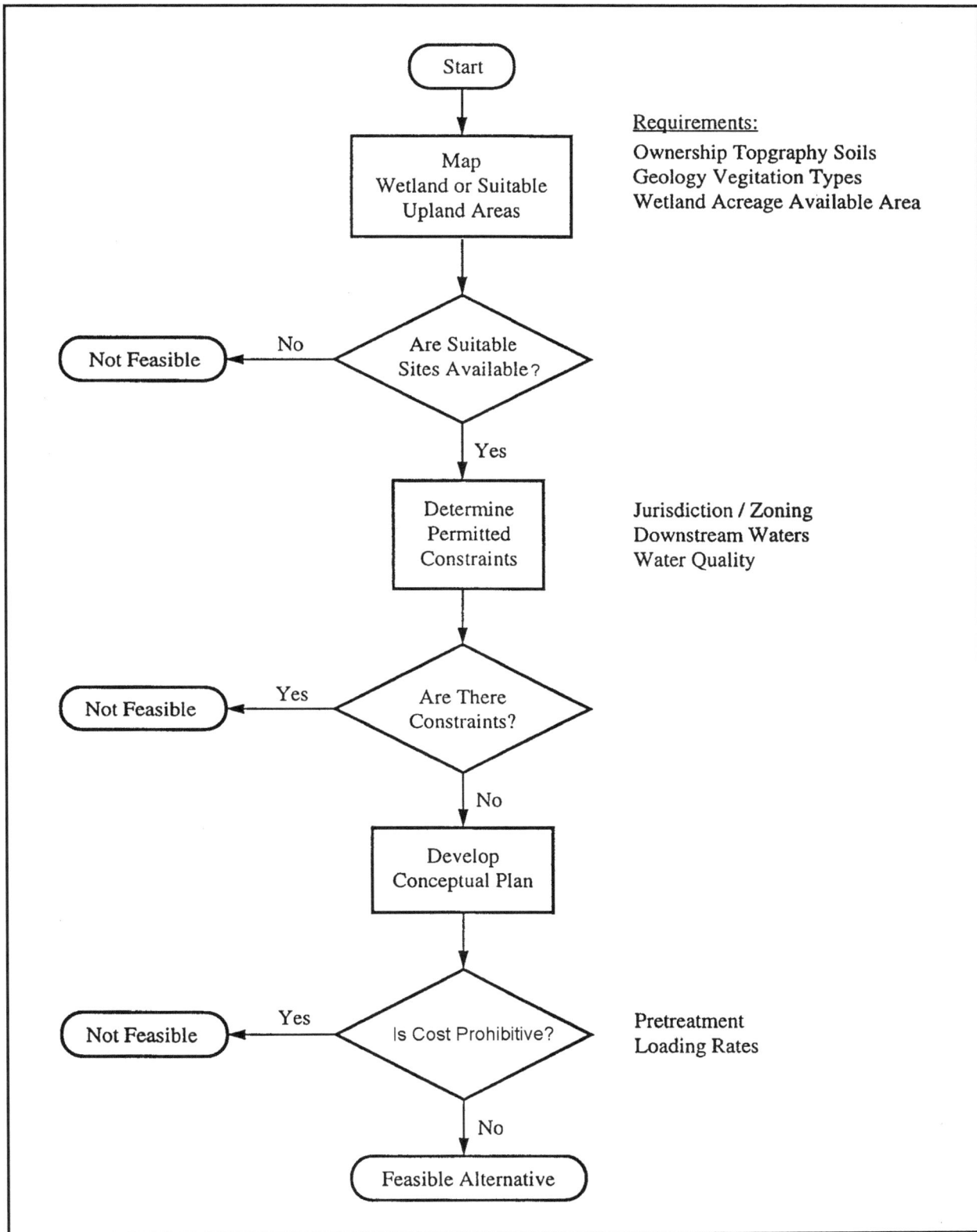

**Figure 5-2. Decision diagram for selecting wetland alternatives (copyright © Water Environment Federation, used with permission)**

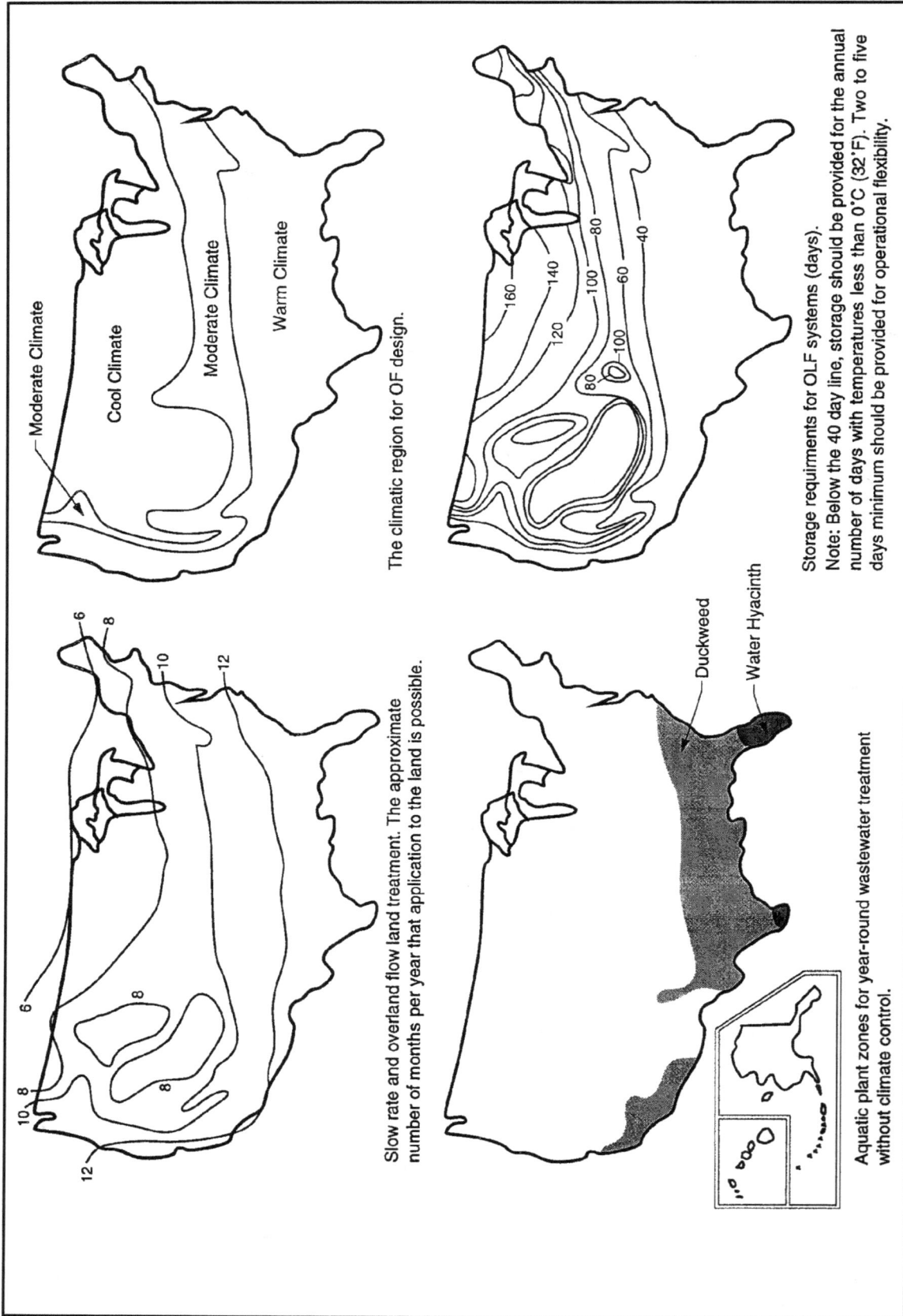

The climatic region for OF design.

Storage requiments for OLF systems (days).
Note: Below the 40 day line, storage should be provided for the annual number of days with temperatures less than 0°C (32°F). Two to five days minimum should be provided for operational flexibility.

Slow rate and overland flow land treatment. The approximate number of months per year that application to the land is possible.

Aquatic plant zones for year-round wastewater treatment without climate control.

Figure 5-3. Climate control and zone considerations for land treatment facilities (copyright © Water Environment Federation, used with permission)

groundwater are important considerations. The SR land treatment design procedure is shown in Figure 5-4 (MOP FD-16). A minimum of primary sedimentation or pretreatment is a prerequisite for successful operation; Imhoff tanks with influent grinder pumps have been successfully employed as pretreatment or preliminary treatment processes. SR design considerations include loading rate, allowable soil permeability, field surface areas, and wastewater storage requirements.

(2) $BOD_5$ removal is accomplished by soil adsorption. Microbe oxidation removal efficiencies invariably range above 90 percent. Total suspended solids are effectively removed by the soil filtration process, with many designs achieving suspended solids removals to 1 mg/L. Nitrogen is removed by crop uptake, denitrification, volatilization of ammonia, and soil matrix storage, with removal efficiencies typically varying between 60 and 90 percent. Pathogen removal is generally excellent. Phosphorous adsorption also readily occurs in soils, and 90 to 99 percent reduction can be expected in both cold and warm climates.

(3) SR site characteristics, typical design features, and expected water quality can be found in EM 1110-2-504, EPA/625/1-84/013a, Metcalf & Eddy 1991, MOP FD-16, and WEF MOP-8.

*c. Rapid infiltration (RI) land treatment systems.*

(1) RI, a well established natural system also known as soil-aquifer treatment (SAT), operates year-round utilizing primary clarification or Imhoff tanks as pretreatment processes. Treatment is achieved mainly by wastewaters percolating vertically downward through permeable soil columns, making RI the most intensive of the natural systems options. Basically, RI operates on a "fill and subside" regime in shallow basins. RI basins typically are intermittently dosed on 1- to 7-day cycles and rested for 6 to 20 days. The rapid infiltration method generally produces a high degree of treatment, although nitrate concentrations of 10 mg/L have been known to reach underlying groundwaters. More intensive soil investigation is required for RI than for either the SR or the OF method.

(2) To be successful, an RI system must be constructed at a site with more than sufficient area of both permeable and well-drained soil to depths that satisfactorily meet treatment objectives. The RI system has the greatest impact on underlying groundwater quality. Extensive data is required on the hydrogeology of the subsurface to include:

- geometry of the system.

- hydraulic loading rate.

- minimum depth to the fluctuating groundwater table.

- slope of the groundwater table.

- depth to the underlying impervious formation.

- hydraulic conductivity of the aquifer soil.

- porosity of the soil.

- elevation of and distance to any stream, river, lake, or wetland water surface.

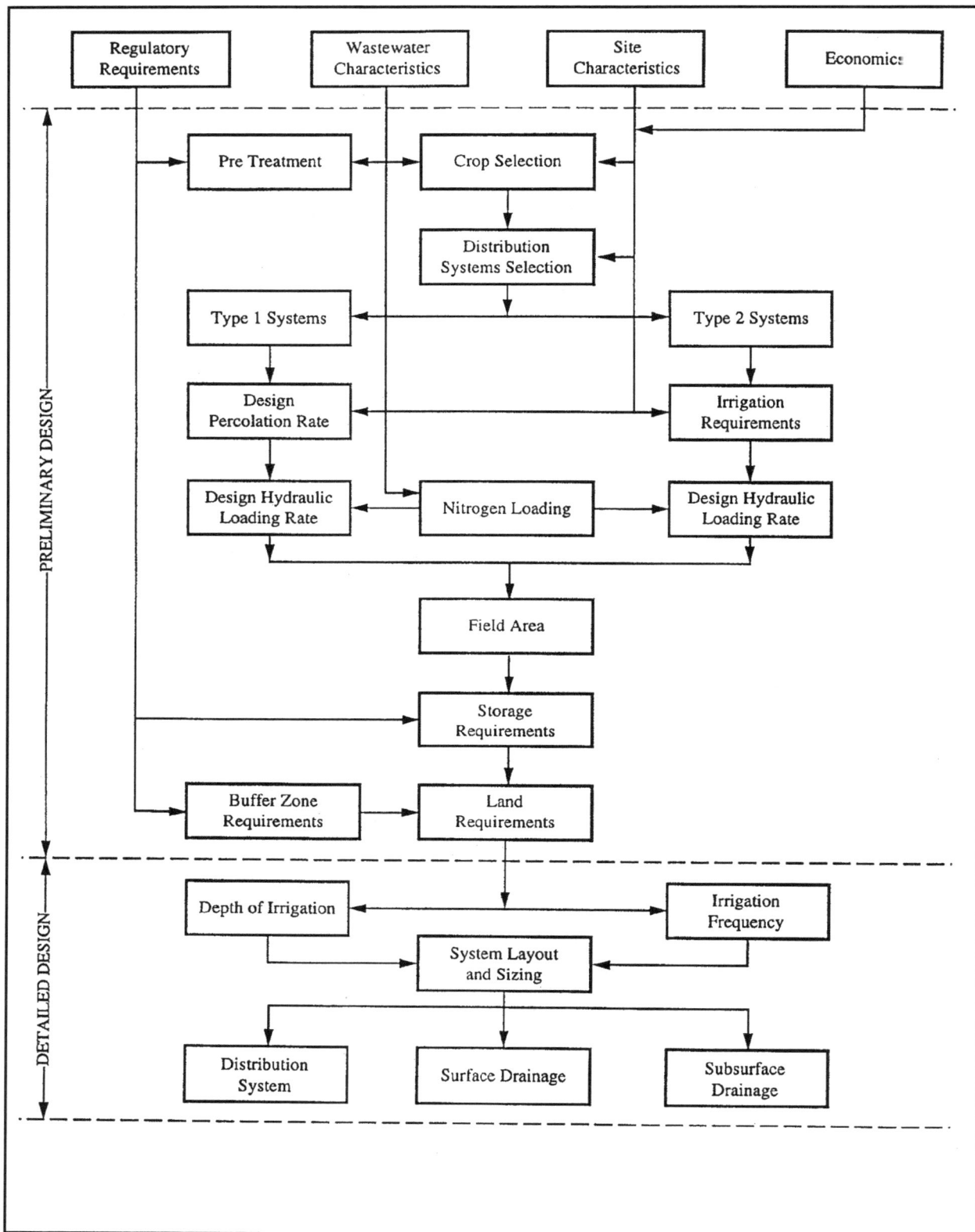

**Figure 5-4.** Slow rate system design procedure (copyright © Water Environment Federation, used with permission)

(3) Loading cycles are seasonally dependent. Hydraulic loading rates with accompanying drying periods for RI systems greatly exceed those of either SR or OF systems.

(4) $BOD_5$ removals of 80 to 100 percent can be expected. Nitrogen removal depends on the raw $BOD_5$ to N ratio, hydraulic loading rate and wet/dry cycle with percentage removals fluctuation over a wide range, i.e., from 10 to 90 percent. Phosphorous removal is dependent upon vertical soil column travel distance. Fecal coliform removal of 2 to 6 log correlate directly with soil composition, vertical soil column travel distance and wastewater application schedules versus bed drying times. Suspended solids removal is high.

(5) Aerobic and facultative ponds are not recommended as pretreatment processes for RI unless algae effluents therefrom are controlled; algae can clog the underlying infiltration surfaces.

(6) Typical design features and considerations for RI systems can be found in EM 1110-2-504, EPA/625/1-84/013a, Metcalf & Eddy 1991, MOP FD-16, and WEF MOP-8.

*d. Overland flow (OF).*

(1) OF is a more recent development than either the RI or the SR process. The system was selected for use in areas with low-permeability, or poorly drained soils which are slowly permeable such as clays and silts. Such conditions necessitate low hydraulic loading rates, thereby requiring a larger application area over a network of vegetated sloping terraces. Wastewater flows down the sloping terraces over the top of the surface with infiltration being limited by the low soil permeability. Wet-dry cycles produce a batch mode treatment, and the treated liquid experiences a variety of physical, chemical, and biological conditions. The combination of sloping terraces 30 to 60 m (100 to 200 ft) at two to eight percent slopes; hydraulic loading rate, wastewater application rate, time of continuous application versus drying periods, application of wet-dry ratio, and time required for a given application cycle in days or hours usually produces a high-quality effluent.

(2) $BOD_5$, TSS, and nitrogen are significantly removed by the OF process; phosphorous and pathogens are removed to a lesser degree. $BOD_5$ effluent concentrations of less than 10 mg/L are often achieved and except for algae, the TSS effluent concentrations are also less than 10 mg/L. Some algae types are not consistently removed. If large algal blooms occur in pretreatment or preliminary treatment processes (e.g., facultative ponds or lagoons), then concentrations of algae in OF algae effluents can be expected to be considerable. Nearly complete nitrification of ammonia can be expected whereas significant nitrogen removal, (i.e., above 80 percent) is difficult to achieve and is dependent on temperature, application rate, and wet/dry cycles. The higher the rate of wastewater application, the quicker the runoff with correspondingly lower treatment efficiency for all parameters.

(3) OF process sheet flow schematics are shown in Figure 5-5 (MOP FD-16 Natural Systems for Wastewater Treatment). Wastewater treatment is achieved mainly by direct percolation and evapotranspiration. Wastewater is generally applied by gated piping at pressures of 14 kPa to 35 kPa (2 to 5 psi), by low-pressure fan spray devices at 35 kPa to 138 kPa (5 to 20 psi), or by high-pressure impact sprinklers at 138 kPa to 522 kPa (20 to 80 psi). The OF process is most successfully operated following a combination of screened wastewater, primary, or pond treatment.

(4) Design considerations for OF can be found in EM 1110-2-504, EPA/625/1-84/013a, EPA/1-81/013, Metcalf & Eddy 1991, MOP FD-16, and WEF MOP-8.

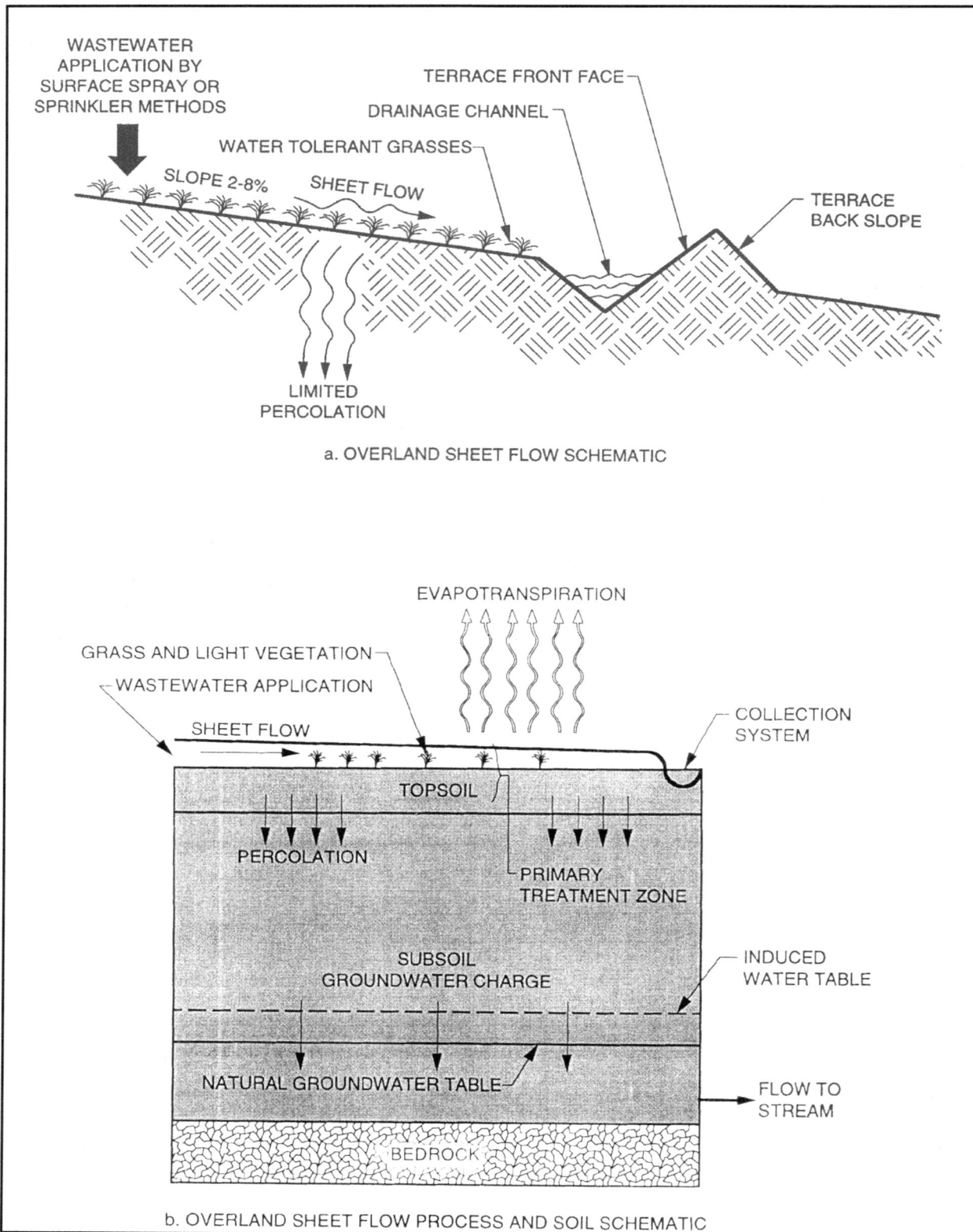

Figure 5-5.  Overland flow process schematic (copyright © Water Environment Federation, used with permission)

## 5-6. Man-Made Wetlands

The Environmental Protection Agency has developed definitions and interpretations to differentiate between natural wetlands and man-made systems (CWA, Section 404), as follows:

  *a. Constructed wetlands.* Constructed wetlands are those intentionally created from non-wetland sites for the sole purpose of wastewater and stormwater treatment. These are not normally considered waters of the U.S. Constructed wetlands are considered treatment systems (i.e., non waters of the U.S.); these systems must be managed and monitored. Upon abandonment, these systems may revert to waters of the U.S. Discharges to constructed wetlands are not regulated under the Clean Water Act. Discharges from constructed wetlands to waters of the U.S. (including natural wetlands) must meet applicable NPDES permit effluent limits and state water quality standards.

  *b. Created wetlands.* Created wetlands are those intentionally created from non-wetland sites to produce or replace natural habitat (e.g., compensatory mitigation projects). These are normally considered waters of the U.S. Created wetlands must be carefully planned, designed, constructed, and monitored. Plans should be reviewed and approved by appropriate state and federal agencies with jurisdiction. Plans should include clear goal statements, proposed construction methods, standards for success, a monitoring program, and a contingency plan in the event success is not achieved within the specified time frame. Site characteristics should be carefully studied, particularly hydrology and soils, during the design phase.

  *c. Natural wetlands.* Natural wetlands have not been fully defined, but certain guidelines and restrictions on use were emphasized. Natural wetlands may not be used for in-stream treatment in lieu of source control/advanced treatment, but may be used for "tertiary" treatment or "polishing" following appropriate source control and/or treatment in a constructed wetland, consistent with the proceeding guidelines.

  *d. Constructed wetlands and wastewater management.*

  (1) Constructed wetlands are areas that are periodically inundated at a frequency and depth sufficient to promote the growth of specific vegetation and are generally categorized as either free water surface systems (FWS) or subsurface flow systems (SFS). Figure 5-6 identifies the types of constructed wetlands most commonly used for wastewater management (MOP FD-16 Natural Systems for Wastewater Treatment). Shallow basins or channels comprise the former with an impervious layer to prevent infiltration plus a supporting vegetative growth medium. Basically, wastewater flows at a low velocity over the medium which supports the growth and through and around the vegetative stalks. Wastewater application is essentially plug flow. The subsurface method is composed of a slightly inclined trench or bed underlaid by an impervious layer to prevent seepage and a permeable medium to support vegetative growth through which the wastewater flows. The root-zone method of rock-reed-filter is categorized as a subsurface flow system.

  (2) The hydrology of wetlands is not significantly different from that of other surface wastewater treatment processes; however, plant growth and substrate do affect flow. Hydrologic factors to be considered in design are the hydroperiod, hydraulic loading rate, hydraulic residence time, infiltrative capacity of the underlying hyperoid, evapotranspiration effects, and the overall water balance. Hydroperiod includes both duration and depth of flooding. Hydraulic residence time of a treatment wetland is the average time a typical unit of water volume exists within the system. Infiltrative capacity is the measure of net water transfer through the sediments either infiltration or exfiltration. Evapotranspiration is the combined water loss from a vegetated surface area via plant transpiration or surface water

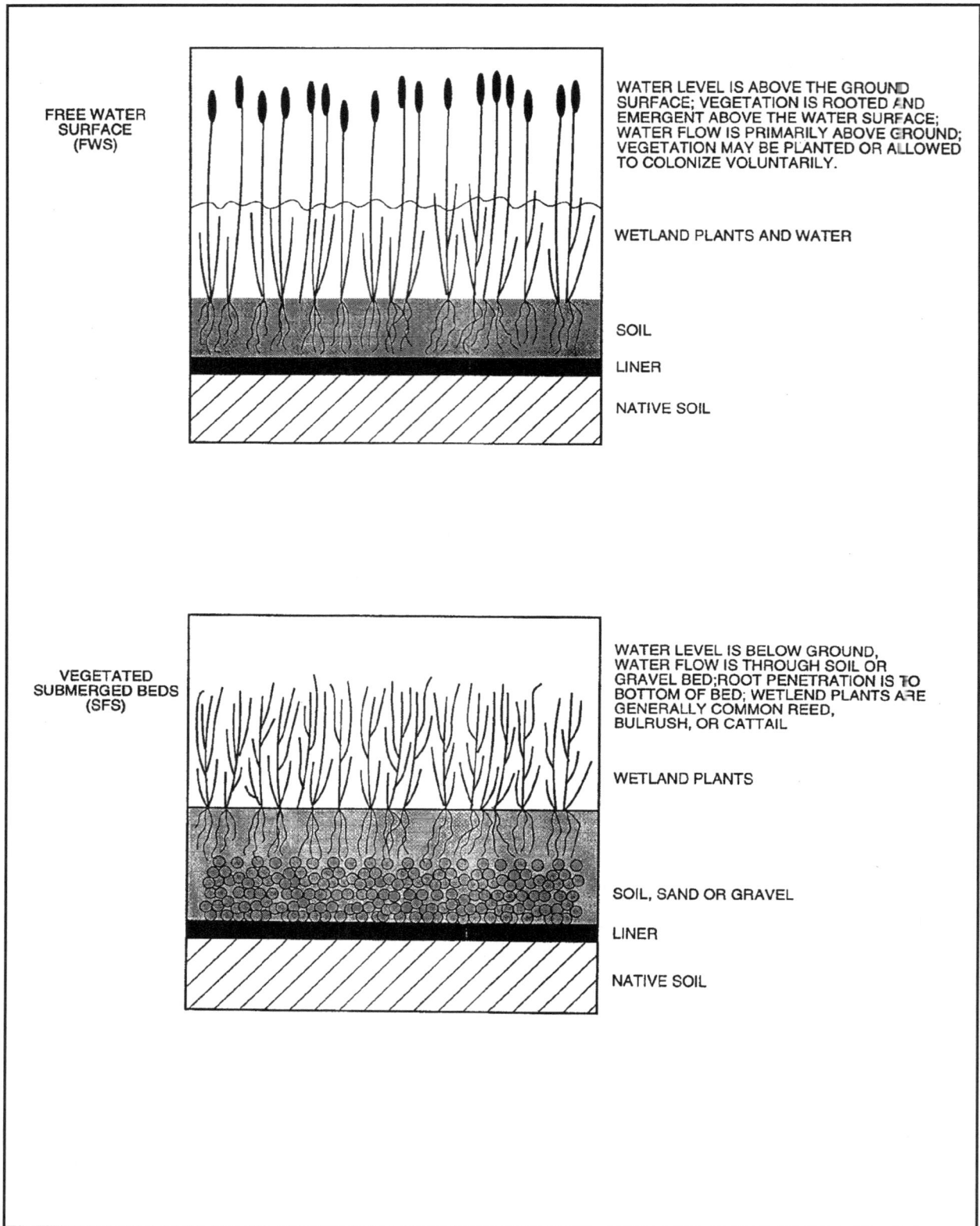

FREE WATER SURFACE (FWS)

WATER LEVEL IS ABOVE THE GROUND SURFACE; VEGETATION IS ROOTED AND EMERGENT ABOVE THE WATER SURFACE; WATER FLOW IS PRIMARILY ABOVE GROUND; VEGETATION MAY BE PLANTED OR ALLOWED TO COLONIZE VOLUNTARILY.

WETLAND PLANTS AND WATER

SOIL
LINER
NATIVE SOIL

VEGETATED SUBMERGED BEDS (SFS)

WATER LEVEL IS BELOW GROUND, WATER FLOW IS THROUGH SOIL OR GRAVEL BED; ROOT PENETRATION IS TO BOTTOM OF BED; WETLEND PLANTS ARE GENERALLY COMMON REED, BULRUSH, OR CATTAIL

WETLAND PLANTS

SOIL, SAND OR GRAVEL
LINER
NATIVE SOIL

Figure 5-6. Types of constructed wetlands (copyright © Water Environment Federation, used with permission)

evaporation. Figure 5-7 identifies the most commonly used effluent distribution methods in wetlands (MOP FD-16 Natural Systems for Wastewater Treatment).

(3) Wetlands have been shown to be excellent assimilators of both $BOD_5$ and TSS. Decomposition is mostly aerobic. Microbial populations apparently flourish similar to activated sludge plants and trickling filter attached growths. $BOD_5$ removal efficiencies are lowest at low input concentrations (less than 5 to 10 mg/L) but increase to 70 to 90 percent removals at higher input concentrations. TSS removal efficiencies are also input related, consistently above 50 and up to 90 percent. Nitrogen removal efficiencies generally are 25 percent or less. Ammonia assimilation by wetlands, while sometimes high, is reduced by short hydraulic loading rates and low temperatures. When other factors do not interfere, total ammonia may be reduced 70 to 90 percent. Total nitrogen removal is also directly affected by loading rates with typical expected removal efficiencies of 75 to 95 percent. Removal efficiencies of phosphorous generally range from 30 to 50 percent. Total phosphorous removal efficiency increases with higher input concentrations and higher hydraulic residence times.

(4) The four basic flow patterns for constructed wetlands are plug flow, step feed, recirculation, and step feed with recirculation in a wraparound pond (more commonly known as "jelly roll"). For more details on wetland design considerations, flow pattern configuration, primary or preliminary treatment requirements, and applicable design data, consult EPA/625/1-91/024, EPA/625/1-88/022, EPA/832/R-93/008, Hammer 1989, NSFC-2, and NSFC-3.

(5) Advantages and disadvantages of siting constructed wastewater treatment wetlands in uplands, on slopes, and in lowlands are summarized in Table 5-1.

*e. Aquatic plants.* Aquatic plants employed in one or more of the natural systems are considered under the generic term of vegetation and are categorized as floating plants or submerged plants. Floating plants (water hyacinths, pennywort, and duckweed) have their photosynthetic portions at or above the water surface with roots extending down into the water column and serving as filtration/-adsorption media for suspended solids and bacteria growth.

(1) Hyacinths. As hyacinths are productive photosynthetic plants, they proliferate and are considered a nuisance in southern U.S. waterways but are highly effective in natural wastewater treatment systems. Nitrogen and phosphorous uptake by hyacinths decreases in colder weather. Therefore, hyacinths are generally most effective south of the 30° parallel in the U.S.

(2) Pennywort. Pennywort is not free floating but intertwines and usually grows horizontally; at high densities the plants experience vertical growth. Pennywort nitrogen and phosphorous uptake is relatively steady in both cold and warm climbs and exceeds the uptake rates of hyacinths in cold weather.

(3) Duckweed. Duckweeds, small freshwater plants, are the smallest and simplest of flowering plants but have one of the fastest reproduction rates of those plants suitable for aquaculture. Duckweed design procedures are generally the same as for facultative lagoons and their effluents often exceed the performance of conventional facultative lagoons for $BOD_5$, total suspended solids, and nitrogen removal. Sometimes the duckweed systems are anaerobic and may require post aeration treatment.

(4) Submerged plants. Submerged aquatic plants are either suspended in a water column or rooted in bottom sediments and are susceptible to being shaded out by algae. They are also sensitive to anaerobic conditions. The full worth of submerged plant systems has yet to be determined.

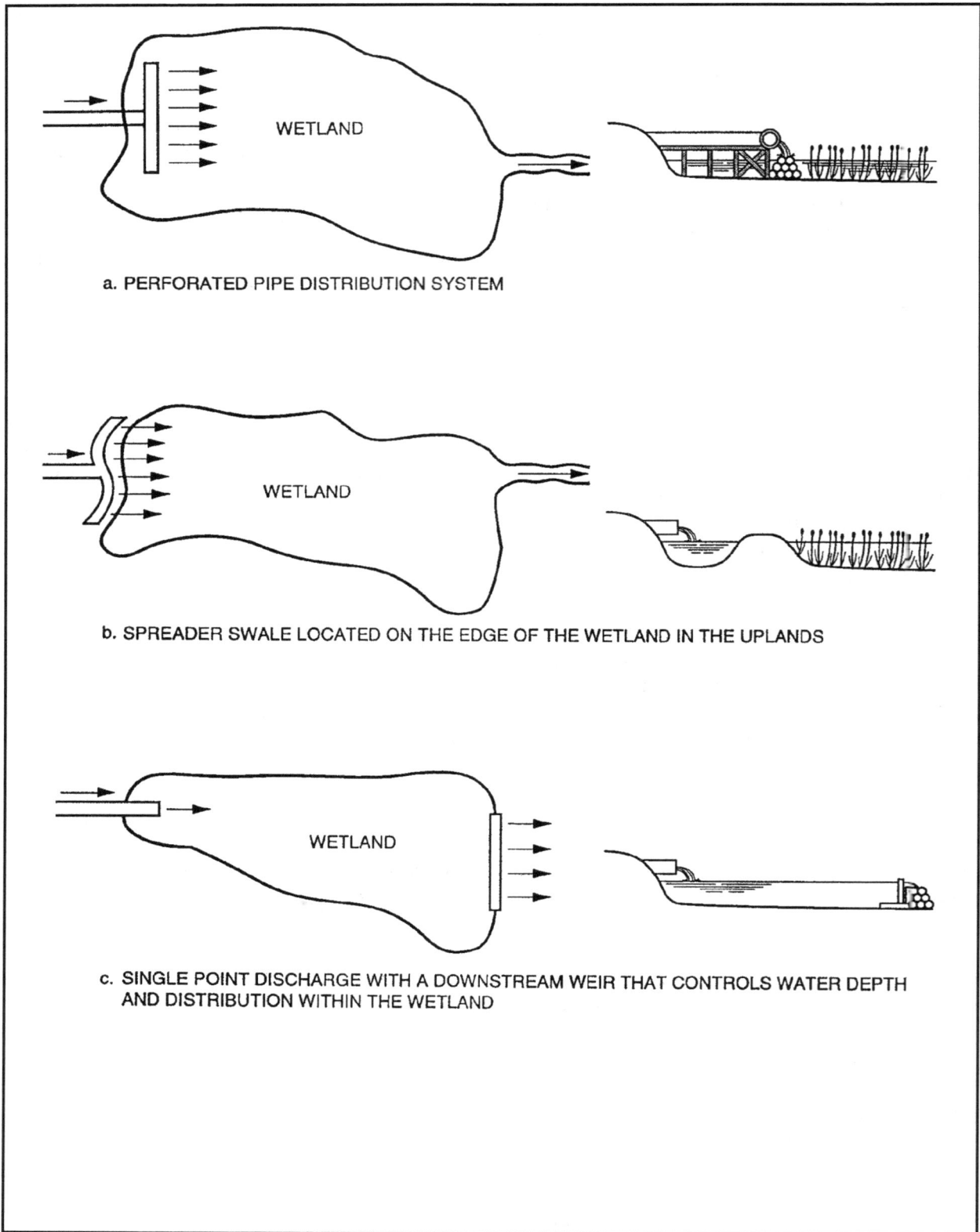

Figure 5-7. Effluent distribution methods in wetlands (copyright © Water Environment Federation, used with permission)

Table 5-1
Advantages and Disadvantages of Siting Constructed Wastewater Treatment Wetlands on Uplands, Slopes, and Lowlands
(Hammer 1989)

|  | Advantages | Disadvantages |
|---|---|---|
| Lowlands | Gravity feed<br>Good construction materials<br>Available dilution water<br>Easy access<br>Well-developed soils<br>Minimal groundwater pollution | Archeological sites<br>Saturated ground<br>High groundwater table<br>Erosion/scouring<br>High flood potential<br>High hydro-period flux<br>T&E, critical habitat<br>Flood plains impact<br>Beavers/muskrats<br>Forest canopy |
| Slopes | Few dwellings<br>Gravity feed | Poor soils<br>Maximum cut and fill<br>Poor access<br>Steep topography<br>Erosion<br>Slope instability |
| Uplands | Minimal cut and fill<br>Minimal flooding<br>Low slope<br>Low erosion potential | Drought susceptible<br>Groundwater pollution<br>No dilution water<br>Extreme hydro-period flux<br>Liners required<br>Shallow bedrock/soils<br>Subsidence potential<br>Maximum downstream use<br>Poor access |

*f. Regulatory concerns.*

(1) Most natural wetlands are classified as waters of the United States. They are also considered state waters in many states. The Clean Water Act prohibits discharges to natural wetlands which do not meet a minimum of secondary treatment or its equivalent (TM 5-814-3). Before discharge into surface receiving bodies (streams, rivers, lakes, impoundments, and estuaries), wasteload allocation processes are often required in order to determine the discharge criteria. Entirely different effluent criteria may be applicable to groundwater discharge. A groundwater monitoring plan will generally be required for all natural treatment systems. Groundwater regulatory standards must be met during the operating lifetime of all natural treatment systems.

(2) Design examples of the three alternative wastewater collection systems can be found in EPA/625/1-91/024.

## 5-7. Nutrient Removal

*a. Conventional and natural treatment.* Nutrient removal of both nitrogen and phosphorous by both conventional and natural treatment systems is of concern as regulatory requirements may impose effluent discharge limits on these two nutrients. In conventional treatment systems both nutrients are generally removed by chemical precipitation. In certain instances, patented treatment processes (such as, Orbal Bionutre, A/O™, A²/O™, Bardenpho™, Modified Bardenpho™, or PhoStrip™) must be used where regulatory standards demand. Each case is site specific; therefore, no trademark nutrient removal processes are included herein.

*b. Nutrients in wastewater.* In considering nutrient removal from wastewater, nitrogen and phosphorous are of prime consideration. Both must be considered potential pollutants in any wastewater discharges.

(1) Nitrogen.

(a) Nitrogen, a ubiquitous element, is a water pollutant in four biochemically interconnectable oxidation states:  organic-N $\rightarrow$ ammonia-N $\rightarrow$ nitrite-N $\rightarrow$ nitrate-N, which taken together compose the total nitrogen in any given sample. As there is almost no nitrate and nitrite to be found in municipal wastewaters, the nitrogen content is most frequently determined as Kjeldahl N, which is the sum of the remaining nitrogenous material, i.e., organic-N + ammonia-N. The behavior of nitrogen in any wastewater system is complex. Organic-N and ammonia-N are oxidized to nitrite and nitrate in wastewater treatment. Nitrites are somewhat unstable and readily oxidize to nitrates. Further oxidation of nitrates will produce inert nitrogen gas ($N_2$).

(b) The conversion of ammonia to nitrate is known as nitrification, a secondary treatment wastewater phenomenon occurring in the presence of the *nitrosomonas* and *nitrobacter* nitrifying bacteria. Conversion of nitrate to nitrogen gas ($N_2$) is known as denitrification, which requires the presence of denitrifying bacteria.

(c) Both nitrification and denitrification require ideal conditions for the most favorable results, and may occur in the same tank, but at different times and in different environments. The principal ingredients required for nitrification and denitrification are sufficient oxygen levels and adequate bacterial concentrations. Currently, in conventional plants, the total nitrogen (organic, ammonia, nitrite, and nitrate) is reduced from influent concentrations of about 40 mg/L to 10 mg/L or less in the discharged effluent. Additional treatment may be required for more stringent effluent quality requirements.

(2) Phosphorous.

(a) The sources of phosphorus are several:  human excreta in water carriage wastes; food residues from households, commercial and recreational establishments and restaurants; water treatment additives; detergents, both domestic and commercial; and indirectly from surface treatment fertilizers through sewer or manhole infiltration. The nutrient phosphorous mainly occurs in solution as particles or waste elements in microorganism forms as: orthophosphates ($PO_4^{3-}$, $HPO_4^{2-}$, $H_2PO_4^-$, and $H_3PO_4^-$); polyphosphates ($P_2O_7^-$); and as organically bound phosphorous, the latter two comprising more than two-thirds of wastewater phosphorous. Phosphorous stimulates the growth of photosynthetic algae, which may cause eutrophication in receiving bodies of water such as lakes or slow-moving streams.

(b) Originally, phosphorous removal in wastewater treatment was achieved by chemical precipitation, principally with ferric chloride. In today's design approaches, treatment is accomplished through the more advanced biological processes. Biological treatment, mainly conventional activated sludge, converts phosphorous to the orthophosphate forms, which are then removed in a chemical precipitation process by calcium, aluminum, or ferric compounds. Normally, if lime is the calcium source of choice, precipitation follows biological treatment. Should alum or iron be employed as precipitants, they may be added in either the primary sedimentation or activated sludge processes.

*c. Combined nitrification-denitrification system.* A two-stage biological nitrogen removal scheme typically employs primary sedimentation, followed by an anaerobic (anoxic) tank for denitrification. The anoxic tank effluent then passes to an aerobic tank for oxidation and nitrification and finally to a secondary sedimentation or clarification tank. Nitrate concentrations are generally returned from the aerobic tank effluent to the anoxic tank influent. Secondary clarifier sludge is either wasted or returned to the anoxic tank. Oxidation ditches (CLRs) with single aerobic and anoxic zones are particularly effective as nitrification-denitrification treatment processes. Nitrogen removals can be significantly reduced further with the addition of a second anoxic zone (Bardenpho process).

   *d. Biological phosphorous removal.* Primary treatment removes about ten percent of municipal influent phosphorous levels. Conventional secondary wastewater treatment processes remove approximately 20 percent of the effluent from the primary process. Typical municipal influent wastewaters have phosphorous concentrations of approximating 10 mg/L, with some regulatory restrictions requiring effluent limits of 1 mg/L or less. Conventional treatment may be inadequate to accomplish the requisite phosphorous removal; chemical precipitation originally became the common phosphorous removal technique when limitations were placed on effluent levels. More recently, certain patented biological phosphorous removal systems have been successfully employed to meet the stringent effluent requirements. As biological processes can reduce primary sedimentation phosphorous levels of applied loads by 70 to 80 percent, a plant influent level of 10 mg/L can be reduced to 2 or 3 mg/L. Phosphorous levels can be further reduced to 0.5 to 1.0 mg/L, if needed, by 3 to 6 mg/L coagulant additions.

   *e. Comparison of nutrient removals by treatment systems.* A brief comparison of nutrient removal processes is shown in Table 5-2. Conventional nutrient removals are identified in the top portion of the table. Certain patented processes are grouped alphabetically in the lower half of the table. The use of patented approaches generally requires license fees for their installation and use. Any license or royalty fees for specific applications are negotiated between the user and the patent holder which ultimately affect any economic analysis of the proposed system.

   *f. Wetland systems.* Wetland systems are particularly efficient in removing one or both wastewater nutrients. The wide range of efficiencies or percent removals is influenced by loading rates, influent constituent strength, soil conditions, plant uptake, and climatic conditions. Removal efficiencies generally increase with higher input concentrations. Table 5-3 identifies the range of variation in nitrogen and phosphorous removals for natural wastewater treatment systems.

## 5-8. Sludge Treatment and Disposal

   *a. Treatment selection criteria.* The selection of sludge treatment methods, if required, is a function of the raw wastewater characteristics, unit processes employed in the small conventional plant treatment train, efficiencies of those treatment processes in removing suspended solids, chemical usages if any (not normally expected in facilities solely treating human wastes), and other site-specific conditions. Sludge disposal is controlled by Federal, state, and local regulations. Chapter 10 summarizes the current national requirements and standards for the use or disposal of sewage sludge.

   *b. Treatment and disposal.* Sludge treatment and disposal depends upon the wastewater treatment selected, as follows:

- For individual units such as pit privies, vaults, compost systems, and septic tanks, collected sludges should be trucked off-site for treatment and disposal.

- For small conventional plants, sludge treatment and disposal depends upon the wastewater treatment process selected but will not necessarily include on-site treatment.

Table 5-2
Comparison of Conventional vs. Patented Processes For Nutrient Removals (EPA/625/1-84/013a)

| Conventional Processes | Capable of Reducing Municipal Wastewater Influents to 20-30 mg/L BOD5 and TSS. | Capable of Producing Plant Effluent of 5 mg/L BOD5 and TSS | Capable of Producing a Plant Effluent of 10 mg/L Nitrate-Nitrogen | Capable of Producing Plant Effluent of 3 mg/L Total Nitrogen | Capable of Producing Plant Effluent of 1.0 mg/L Total Phosphorous | Capable of Producing Plant Effluent of 0.5 mg/L Total Phosphorous |
|---|---|---|---|---|---|---|
| Trickling Filter | Yes | | | | | |
| Activated Sludge | Yes | Yes[1] | | | | |
| Operationally Modified Activated Sludge | Yes | Yes[1] | Yes[1] | | Yes[1] | |
| Oxidation Ditches (CLR) | Yes | Yes[1] | Yes[1] | | | |
| Sequencing Batch Reactors | Yes | Yes[1] | Yes | Yes[1] | Yes[1] | |
| **Patented Processes** | | | | | | |
| A/O | Yes | Yes[1] | | | Yes[1] | |
| A2/O | Yes | Yes[1] | Yes | | Yes[1] | |
| Bardenpho | Yes | Yes[1] | | Yes[1] | | |
| Modified Bardenpho | Yes | Yes[1] | | Yes[1] | Yes[1] | |
| Bionutre | Yes | Yes[1] | Yes | Yes[1] | Yes[1] | |
| PhoStrip | Yes | Yes[1] | Yes[1] | | Yes | |
| Additional Chemical Treatment | | | | | | |
| Chemical addition of alum, lime, or iron salts | | | | | | Yes |

[1] Process should be capable of meeting indicated standard with proper design, acceptable influentwastewater characteristics, and/or tertiary filtration.

**Table 5-3**
**Variations in Nutrient Removals by Natural Wastewater Treatment Systems (NSFC-2)**

| Natural System | Nitrogen Removal | Phosphorous Removal |
|---|---|---|
| Floating Plant Systems | 10-90% | 30-50% |
| Stabilization Ponds | 40-90% | 60-80% |
| Slow Rate Systems | 60-90% | >90% |
| Rapid Infiltration Systems | 10-90% | 30-99% |
| Overland Flow Systems | up to 80% | 20-60% |
| Subsurface Systems | 10-40% | 85-90% |
| Wetland Systems | 75-90% | 30-50% |

*c. Transport.* The USDA Forest Service has successfully managed sludges generated at domestic wastewater treatment facilities at remote sites (USDA-2). These include the use of 210-L (55-gal) drums to store the sludge. When drums are filled, they may be transported to an offsite treatment facility by truck, helicopter, mule/horse pack, or all terrain vehicle (ATV) (USDA-2).

*d. Stabilization.* Sludge stabilization is provided to eliminate nuisances and reduce health-related threats. In small plants, stabilization depends upon whether the sludge is anaerobically or aerobically digested in the treatment train, whether chemical stabilization is employed, and whether composting of sludge is attempted for disposal purposes. Usually, in small conventional plants, sludges are not gravity- or polymer-thickened, nor are mechanical means (centrifuges, filter presses, horizontal belt filters, or rotary vacuum filters) applied.

*e. Lagoons.* Infrequently, at small plants some sludges are lagooned. The limitations on using lagoons as biological digesters are the availability of appropriate land areas, aesthetic or odor problems, lack of flexibility in operations, and generally poor performance in humid and rainy climates. Often, several years may be required for lagoon sludge drying to complete a life cycle. Lagoons must be diked to prevent surface runoff from entering. Lagoons may be used in emergency operations and may be placed in operation after the plant has been constructed, should the need arise.

*f. Dewatering and sand drying beds.* Dewatering of anaerobically or aerobically treated sludge reduces the amount of water in the sludge so that it can be handled and disposed of as a solid rather than a liquid. The simplest dewatering method for small treatment facilities is the use of uncovered or open sand filtration drying beds. The use of sand drying beds, which may be enclosed for aesthetic reasons, is applicable for smaller conventional treatment plants if the residues are well stabilized. Design considerations include: solids concentrations of the applied sludges; proposed depth of sludge to be applied to the bed; efficiency of the filtered water collection system; degree, type, and sludge conditioning in the treatment processes; climatically affected evaporation rate; method to remove dried sludge cake from the beds; and ultimate sludge disposal method contemplated.

(1) Drying beds are usually 6 to 9 m (20 to 30 ft) wide and 7.5 to 38 m (25 to 125 ft) long, and consist of 150 mm to 250 mm (6 in to 10 in) coarse sand layers in rectangular reinforced concrete-walled enclosures over a 150 mm to 300 mm (6 in to 12 in) graded gravel layer. The bottom of the bed should be impervious. A minimum of two beds should be provided for each plant. The sand media subgrade is sloped and a manifold-lateral system of perforated pipe is installed to collect the liquid, leaving the solid concentrations on top of the sand layer. Evaporation of surface waters to the atmosphere aids in the

dewatering process. Filtrate from the manifold-lateral collection system should be returned by pumping to the head of the plant.

(2) Splash blocks at gate locations in the concrete walls and concrete pads for truck or front-end loader wheels should be provided. Sludge is removed either manually or by careful loader scraping to remove as little of the sand layer as possible in each emptying action. Sand replacement is inevitable after several years of operation. Trucked sludge is directed to a local municipal landfill or sludge monofill. Bed-dried sludge is removed by contract at most government plants.

(3) As an operator of over 100 small wastewater plants, the U.S. Army has an interest in efficient and cost-effective sludge dewatering systems. The majority of currently operating Army plants use conventional sand-drying beds to dewater sludges. The U.S. Army Engineer Construction Engineering Research Laboratory (USACERL) has recently evaluated the Reed bed technology for potential Army use (USACERL 1993). Reed bed dewatering is a relatively new modification to sand drying beds which uses a common reed (genus *Phragmites*) to treat wastewater sludges. Sludge is applied to an actively growing stand of reeds. The sludge dries naturally in the sand beds by evaporation and drainage. This technology has been successfully demonstrated in the northeastern United States, where some 50 sludge drying beds are in operation. Reed beds are easier to operate and maintain than regular sand drying beds and virtually eliminate the need for regular sludge removal.

(4) Sludge treatment and disposal design information can be found in Table D-3.

## 5-9. Disinfection of Wastewater Effluents

*a. General.* The most common means of wastewater disinfection is by liquefied chlorine gas or in the form of varying chlorine compounds. Disinfection is defined here as inactivation of all microorganisms. Disinfection in one form or another is almost always required to meet National Pollutant Discharge Elimination System permit requirements or state-imposed effluent quality standards. Concerns for side-effects and/or undesirable chlorine by-product formation, as well as toxic chlorine effect on humans, aquatic life, and shell-fish growing regimes have increased the need for carefully controlled dechlorination practices or alternative disinfectant processes for some discharges.

*b. Non-chlorine techniques.* Recent technology has emphasized other techniques for treating wastewater discharge effluents, particularly ozonation and ultraviolet light. Other available techniques include several chemical disinfectants and their compounds such as bromine chloride, iodine, hydrogen peroxide, and potassium permanganate. Thermal and radiation processes have not been actively pursued. Of the non-chlorine techniques, ozonation and ultraviolet light appear most promising.

*c. Chlorination.*

(1) Chlorination is the most widely used method of disinfection and is accomplished with gaseous chlorine, hypochlorites, or chlorine dioxide. Chlorine demand to inactivate organisms is the difference between applied chlorine and the residual chlorine and provides a measure of disinfection and all other chlorine-demanding reactions for the contact period between the disinfectant's injection into the wastewater and effluent release. There are numerous configurations of chlorine control systems, all requiring special delivery and control equipment, weighing scales (for accurate measurement of dose rates), and carefully controlled dosage devices. For small treatment plants, liquified chlorine gas is delivered in pressurized containers--usually 45.5-kg (100-lb) and 68-kg (150-lb) cylinders. Should dechlorination of effluents be

required to meet effluent quality standards, specific feed rates of sulfur dioxide, sodium sulfite, sodium bisulfite, and sodium thiosulfate have been developed for use in either in-line or contact mixing to nullify the effluent residual chlorine concentrations prior to discharge to the receiving body of water.

(2) The design of any attendant chlorination facility at a wastewater plant must provide automatically controlled forced venting of chlorinator and chlorine tank storage rooms, and a chlorine contact chamber with a detention period of not less than 30 min following chlorine injection. An inspection window or glass pane door panel must be provided to permit outside viewing of the chlorinator installation. Both chlorinator and chlorine tank storage rooms must be provided with a means of heating to at least 16°C (60°F).

(3) Chlorinator equipment must have the capacity to provide design dosage requirements at maximum anticipated plant flow. A solution-feed chlorinator, either pressure feed or vacuum feed, will be required with a suitable make-up water supply. Scales of sufficient size will be required to meet maximum effluent flow rates; preferably the scales will be self-reading to record continuous chlorine application. Piping will be that approved by the equipment manufacturer and will be color coded throughout the treatment facility.

(4) Hypochlorinators, if employed in small flow plants of 75 700 L/d (20,000 gal/d) or less, must be of the positive-displacement metering type and located in a separate room. The hypochlorinator solution feed is chiefly made from commercially available calcium and sodium hypochlorite salts in both dry and liquid form, respectively. The most popular form of hypochlorite is of the sodium type, the liquid form of which can be prepared onsite and commercially delivered at 12 to 15 percent available chlorine. Hypochlorites, in terms of available chlorine, have the same oxidizing potential as chlorine gas. With the exception of the liquid feeder, storage, and piping, a hypochlorination system is quite similar to larger chlorine gas systems. However, since the chlorine is not in the gaseous form, there is less concern with chlorine gas leaks and exposure of personnel.

(5) Design considerations for disinfection by chlorination can be found in Droste 1997 and Metcalf & Eddy 1991.

*d. Ozonation.* Ozone with its high oxidation potential has been used for many years, particularly in Europe, and has received growing attention as a chlorine alternative for disinfection. Ozone may be superior in its ability to inactivate viruses and is equally able to inactivate bacteria. Ozone reduces wastewater odor, produces no dissolved solids, is not affected by pH, and increases oxygenated effluents. As an unstable gas, ozone quickly breaks down into elemental oxygen. Ozone is generated on-site by commercially available generator equipment. Design considerations for the equipment installation include on-site electricity availability, and automatic devices to control voltage, frequency, gas flow, and moisture. The primary deterrents to the use of ozone at small treatment systems are relatively high capital and operating costs.

*e. Ultraviolet light.*

(1) Disinfection by ultraviolet (UV) light is increasingly used as a wastewater effluent disinfection process at both military and municipal facilities. UV disinfection is a process by which ultraviolet light is used to destroy pathogenic organisms. Absorption of UV light by microbes is believed to cause damage of cellular DNA and protein. As cellular damage is caused by UV light, no residuals are added. Germ-killing (germicidal) UV light can be generated by low-pressure mercury lamps. The process has been shown to be a viable, reliable, safe, and cost-effective alternative to the chlorination/dechlorination method. While there

are some disadvantages to using UV, the advantages appear to outweigh any negative implications, as shown in Table 5-4.

**Table 5-4**
**Advantages and Disadvantages of Ultraviolet Disinfection**

| Advantages | Disadvantages |
|---|---|
| Excellent performance | Potential re-activation of irradiated organisms |
| Short contact times | Limited time-span of performance as an accepted technique |
| No undesirable by-products | Uncertain accuracy of reliability in measuring UV dose |
| No required chemical additions | Frequent/expensive apparatus maintenance |
| No chemical changes to effluent | Treatment efficiency not immediately determinable |
| No physical changes to effluent | |
| No potential harm to downstream humans, fish-life, or aquatic plants | |
| No detrimental overdosing effects | |
| Competitive costs | |
| No disinfectant residual in effluent | |

(2) A typical UV disinfection unit consists of UV lamp modules placed in an open reactor or a wastewater channel. Depending on the UV intensity requirements, two or more UV lamps are placed adjacent to each other or at a predetermined distance to or from individual modules. Lamp modules may be configured horizontally or vertically.

(3) For a bank, two or more lamp modules may be placed adjacent to each other across the width of the channel or a reactor. UV banks may be placed in series or in parallel depending on space availability and permissible head losses. Reactors may be made of concrete, metal, or UV-resistant plastic. Reactor dimensions vary depending on the flow and the size and arrangement of the lamps and the desired retention time. Flow control devices such as weirs or automatic level control gates are provided to maintain steady state flow and constant water level. UV reactors can contain vertical or horizontal banks placed in series or in parallel depending upon flow, microbial population, and percent inactivation required.

(4) UV disinfection units usually include control panels and power distribution panels. Configurations and features of power distribution panels and control panels vary widely depending on manufacturer specifications and models.

(5) Design considerations for UV disinfection include: delivered UV dose to provide the required inactivation (UV dosage is a function of UV intensity as well as retention time within the banks of the lamps); hydraulic characteristics of the reactor which must closely resemble turbulent plug flow; flow pacing which is not a concern in small plants; reactor hydraulic level control; availability of critical chamber head loss; wastewater characteristics including suspended solids, $BOD_5$, dissolved organics, and color; and redundant UV disinfection units if required by state or local regulations. Also of importance are individual lamp condition, ultraviolet light intensity, and projected lamp life.

(6) Small wastewater treatment plants are especially adaptable to the UV disinfection process for up to 379 000 L/d (100,000 gal/d). UV disinfection units may be designed to meet specific disinfection goals, or prefabricated submersible units may be used and retrofitted at existing plants.

(7) In the design of a UV disinfection system, the following steps must be performed:

Step 1—Collect wastewater data.

Step 2—Establish mathematical model coefficients and parameters.

Step 3—Establish reactor parameters and equipment conditions.

Step 4—Determine reactor UV density.

Step 5—Establish UV radiation intensity.

Step 6—Establish inactivation rates.

Step 7—Set hydraulic rates.

Step 8—Establish UV loading/performance relationship.

Step 9—Establish performance goals.

Step 10—Determine reactor size.

(b) Design considerations for disinfection by UV light can be found in ETL 1110-3-442.

# Chapter 6
# Laboratory Design, Sampling, and Flow Monitoring

## 6-1. General

The purpose of a well-equipped and properly operated laboratory is to control and monitor the operation of the wastewater treatment facility. Sampling and flow monitoring facilitate compliance with regulatory monitoring requirements and process control. For small-scale treatment facilities, however, on-site laboratory facilities may not be cost effective. The decision to construct on-site laboratory facilities is primarily economic. Commercial laboratories, as well as centrally located Engineer Division or District laboratories, may prove to be more cost effective. For a detailed description of responsibilities, policies, materials and chemistry testing, and analytical services capabilities of the major subordinate commands (MSC) laboratories, see ER 1110-1-8100. In addition, ER 1110-1-261 describes responsibilities and procedures for laboratory testing performed by and for the Corps of Engineers District Offices.

## 6-2. Laboratory Design

If included as a facility feature, the laboratory should be located on the ground floor or in the basement, preferably with a northerly exposure to light; the laboratory should have a solid floor and should be free of traffic and machinery vibrations.

 *a. Space.* The first criterion with regard to laboratory facilities is the floor space required. Generally, floor space is based on square meters (square feet) per person working in the laboratory. Also necessary to consider are storage space, office space, and special areas dedicated to testing for specific parameters, as well as space for the installation of hoods, benches, cleanup stations, etc. In general, the flow of work in the laboratory should be considered in lab bench and equipment layout arrangement, with a minimum of people working in the same area at the same time. "Quiet" areas may need to be provided for some work assignments. Some states require a minimum laboratory square footage, particularly for bacteriological examinations. Each laboratory should comply with Federal OSHA regulations. Guidance on design of laboratories is available in USAEPA-2.

 *b. Materials.* Acoustical tiles should be used for ceilings. Light colors are recommended for all interior walls. Floors should be either vinyl or rubber tile, and fire resistant as well as resistant to acids, alkalies, solvents, and salts. Doors should permit straight egress from the laboratory, and should have glass windows. All metals used in the construction of cabinets should be U.S. standard gauge 18 or better. All sheet metal should be coated with a corrosion-resistant finish. The shelf surface should be a smooth, hard, satin luster resistant to acids, alkalies, solvents, abrasives, and water. Stainless steel should be ANSI type 316 (OSHA 1996).

 *c. Utilities.* The laboratory should be supplied with water, gas, air, and vacuum service lines and fixtures; traps, strainers, overflows, and plugs. Electrical service outlet fixtures should be convenient and adequate, preferably located on all laboratory walls.

 *d. Sinks.* Generally, the laboratory should have one double-well sink with drain board. Sinks should be made of epoxy resin or plastic material with all appropriate appurtenances for laboratory applications. Water fixtures on which hoses may be used should be equipped with reduced-zone pressure backflow devices to prevent contamination of the water lines. Sinks should be highly resistant to acids, alkalies, solvents, salts, abrasives, and heat. Traps should be easily accessible.

*e. Ventilation and lighting.* Laboratories should be separately air conditioned, with external air supply providing 100 percent makeup volume. Separate exhaust ventilation should also be provided. Good lighting is also essential.

*f. Power.* To prevent line fluctuation, all electrical lines coming into the laboratory should be controlled with CVS harmonic neutralized-type transformers. For higher voltage requirements, the 220-volt lines should be regulated accordingly.

*g. Gas.* Natural gas should be supplied to the laboratory. Gas outlets should be placed in readily accessible locations.

*h. Laboratory equipment.* For minimum laboratory equipment requirements, see TM 5-814-3, Appendix F.

## 6-3. Sampling and Analysis

Proper operation of a wastewater treatment facility depends upon a well-defined and site-specific sampling and analysis program to monitor the performance of the treatment processes and ensure compliance with the regulatory requirements. For specific and general guidelines on sampling and analysis programs, see TM 5-814-3, Chapter 18. In addition, EM 200-1-3 provides guidance for preparing project-specific sampling and analysis plans (SAPs) for the collection of environmental data. For additional laboratory safety and health requirements, consult EM 385-1-1 and WEF MOP-1.

## 6-4. Flow Monitoring

*a. Measurement devices.* In conventional wastewater treatment plants, flow measurement is probably the most important element in collecting plant monitoring data. Primary flow measurement devices produce a hydraulic transition between subcritical and super-critical flow by resting the channel. Sharp-crested, broad-crested, V-notch, and proportional weirs can be used to measure flows accurately, but they tend to trap sediment and pollutants. Flumes are generally not affected by sedimentation or obstruction problems, particularly if they are located downstream of bar racks or other coarse screening devices. The most popular devices are Parshall flumes for open-channel applications and Palmer-Bowlus flumes for in-pipe flow measurement. Parshall flumes, the most often used wastewater flow measuring devices, have a lower head loss than a weir and a smooth hydraulic flow which prevents deposition of solids. Generally, Parshall flumes may be purchased in pre-fabricated forms with the necessary sonic or other measuring devices and recorders, which read both instantaneous flows and totalized daily flows.

*b. Design formulas.* Design formulas and tables for weirs, flumes, and flow monitoring equipment can be found in TM 5-814-3.

## Chapter 7
## Treatment Process Selection

### 7-1. Overview

It is very important that safe and effective disposal of human and domestic wastes be provided in recreational areas to ensure the preservation of the quality of surface water and groundwater. The selection of appropriate wastewater treatment facilities for recreational areas should be based on site visitation, design considerations, local resources, economics, health factors, aesthetics, safety, and access. A discussion of these parameters and how they affect the selection of the treatment process is presented in this chapter. Figure 7-1 presents the typical wastewater treatment and disposal alternatives available for treating waste produced at USACE recreational areas, and compares of the advantages/capabilities and disadvantages/limitations of these processes.

### 7-2. Site Visitation

For recreational facilities with less than 30,000 visits each year, the design engineer may consider selecting a wastewater collection system that does not involve water-carried waste. These would be single-unit installations such as comfort stations and facilities in remote areas. Generally, soil, climate, and availability of water and power will dictate the selection for this type of facility.

### 7-3. Local Resources

*a. Resource-limited sites.* Certain sites may be resource limited and may require specialized systems. For example, a comfort station having minimal quantities of water may require a plan using a combination of water for hand washing and a non-potable water source unit for urinals and water closets. Such a design would allow for the segregation of graywaters and blackwaters and possibly simplify the overall system design. Gray wastewater may, in some instances, be treated onsite by septic tanks and absorption fields. In other instances, it may be necessary to include additional facilities for pumping and trucking wastewater to a central facility for further processing.

*b. Other sites.* Other sites may not be resource limited and, when the annual visitation is small, may allow a total on-site treatment of wastewater through utilization of the appropriate processes.

### 7-4. Economic Considerations

Economic considerations must be site-specific and based upon alternatives available for a particular site.

*a. Ranking of treatment alternatives.* Computer-assisted techniques can be used to rank different wastewater treatment alternatives, each capable of meeting specified effluent criteria, on the basis of cost effectiveness. Two currently available computer programs which can aid the design engineer in the design and selection of recreational treatment facilities are described below. Both programs rank different alternatives based on overall cost estimates including capital costs and operation and maintenance (O&M) costs.

*b. Capital costs.* Capital costs are those associated with the purchase of land, equipment, plant construction, and other related facilities. The most accurate capital cost can be estimated by obtaining price

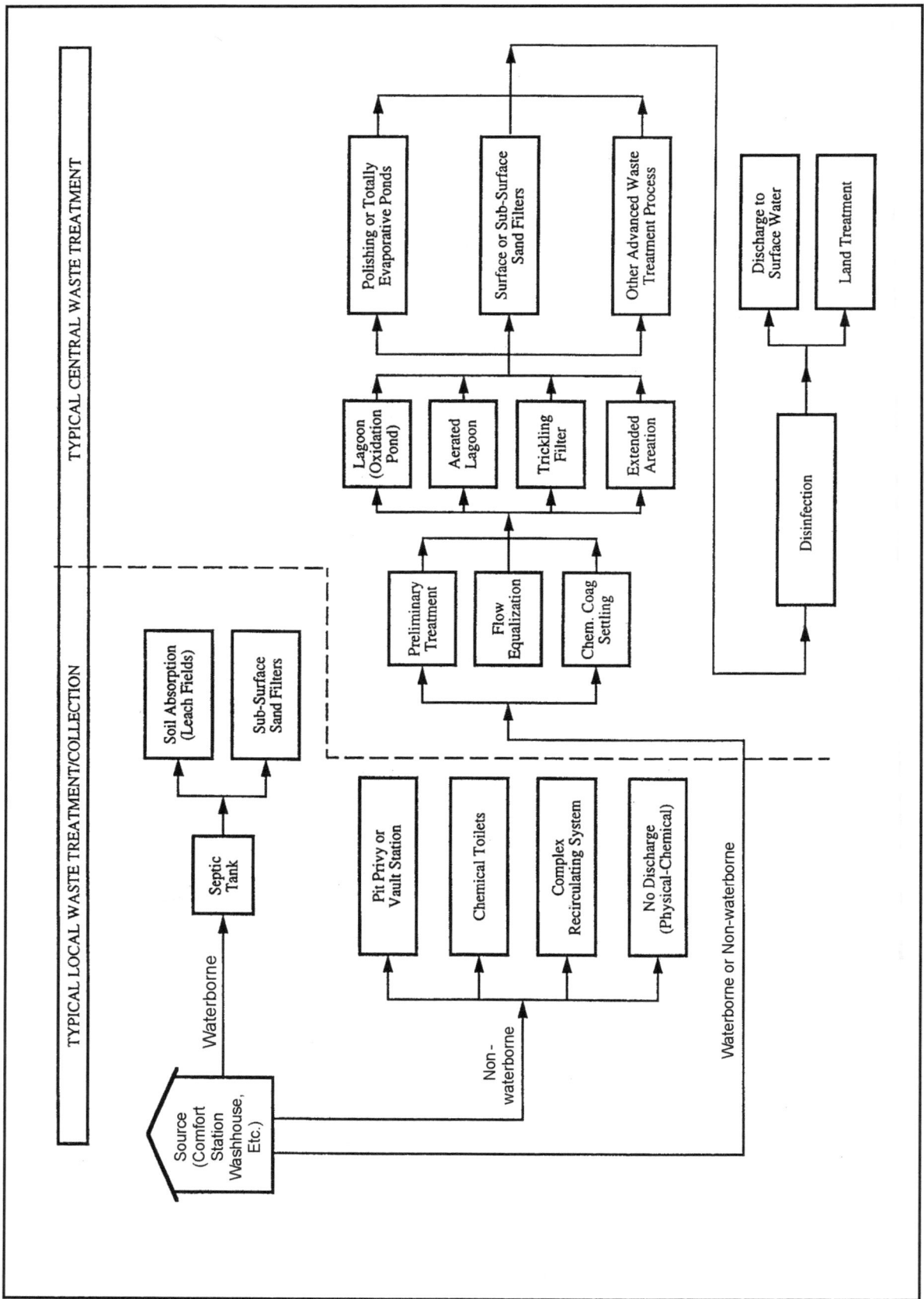

Figure 7-1. Typical wastewater treatment and disposal alternatives available for CE recreation areas

quotes from local equipment suppliers and contractors. If time and budgetary constraints prohibit the design engineer from obtaining actual quotes, the capital cost of any size treatment system may be estimated based on past costs. Because costs continually change, it is important that the capital costs of any treatment altenative are referenced to the same cost indices. The Environmental Protection Agency (EPA) periodically publishes wastewater treatment plant and sanitary sewer cost indices. One of the most commonly used cost indices is the Engineering News-Record Construction Cost (ENRCC) Index.

*c. O&M costs.* O&M costs are annual costs and for most treatment processes include the following categories: labor (supervision, report preparation, clerical, laboratory, yard, operation, and maintenance), power, chemicals, parts, supplies, and monitoring.

*d. CAPDET.* The Computer Assisted Procedure for the Design and Evaluation of Wastewater Treatment Systems (CAPDET) was developed to provide accurate planning-level cost estimates. CAPDET has a component that specifically addresses small systems (flows less than or equal to 3 785 000 L/d (1,000,000 gal/d)), and includes programs to adjust unit labor, chemical, and other prices for current market conditions. It is strongly recommended that prior to using CAPDET, the design engineer becomes familiar with this program and, more importantly, with its limitations. The program is available from Hydromantis, Inc., 1685 Main St. West, Suite 302, Hamilton, Ontario, Canada L85 IG5 (Hydromantis 1992).

*e. ECONPACK.* PC-ECONPACK is a comprehensive economic analysis computer program that incorporates economic analysis calculations, documentation, and reporting capabilities. This program was developed to comply with the regulations governing proposed military construction projects within the Department of Defense. These regulations require that each construction request project estimate for facilities investments be accompanied by an economic analysis. ECONPAK performs standardized life-cycle cost calculations such as net present value, equivalent uniform annual cost, savings-to-investment ratio, and discounted payback period. More information on ECONPAK can be found in USACE 1986.

## 7-5. Health Considerations

*a. General.* As a general principle, waterside recreational treatment facilities should be located along a section of the receiving body of water having a low mosquito production potential. The normal summer water-level fluctuation zone should be identified and completely cleared of vegetation. Vegetation of a type and density favorable to mosquito production in flat, protected areas within the normal summer fluctuation zone should be periodically controlled by mechanical or chemical measures. Regulation of the water level in stabilization ponds and other man-made impoundments is an effective means of controlling aquatic weeds near dikes.

*b. Mosquito control.*

(1) In many U.S. locations, breeding of mosquitoes in natural and constructed wetland treatment systems may ultimately determine treatment system selection. Prevention of disease transmission and the suppression of mosquito-borne nuisance levels must become an objective of mosquito control techniques in any treatment environment. Often fish populations (particularly *Gambusia* spp.) are bred to control mosquitoes; however, fish cannot tolerate the anaerobic conditions when ponds stagnate or become organically overloaded. Thus, if plant growth conditions become dense, say in hyacinth systems, mosquitoes may develop and flourish. Also, some areas of such systems may be accessible to the multiplying mosquitoes but not to the fish.

(2) In addition to stocking ponds with fish, mosquito control strategies include:

- more effective pretreatment to reduce total organic loading on the aquatic system, thereby maintaining aerobic conditions.

- step-feed of influent waste stream with recycle.

- more frequent plant harvesting.

- water spraying in the evening hours.

- application of chemical control agents (larvicides).

- diffusion of oxygen with aeration equipment.

- biological control agents (e.g., BT/*israelensis*) (Metcalf & Eddy 1991).

(3) Provisions should be made for proper storage, collection, and disposal of garbage and refuse throughout all recreational areas in order to prevent and control flies. Care should be taken to ensure that screenings, etc., from wastewater treatment facilities are adequately protected from, and inaccessible to, flies.

## 7-6. Aesthetic Considerations

It is essential that wastewater treatment facilities not encroach upon the natural, scenic, aesthetic, scientific, or historical value of the recreational area. For maximum benefits to be derived from a recreational area, these facilities be designed using sound engineering principles and aesthetic judgment as well. The design engineer must ensure that recreational treatment facilities are located well away from the recreational area, and that land treatment systems and waste stabilization ponds are located downwind from the recreational treatment facilities. Odors can be controlled with masking agents or by using chemical additives (Ehlers 1965). When odors are associated with pumpage from septic tanks, it is best to pump and transfer wastes when the recreational area is closed or visitation is at a minimum. Preplanning conferences, open to all interested parties and agencies, should be held to assist planners in ensuring that the recreational wastewater treatment facility can serve the needs of the recreational area without impairing its future use.

## 7-7. Safety Considerations

The design engineer has the responsibility of incorporating as many safety features as possible into the plant design, the plant grounds, and all ancillary operations such as collection systems, lift stations, effluent structure, and standby generators. For specific safety requirements and their implementation, consult Occupational Safety and Health Administration (OSHA) standards and applicable Army regulations.

## 7-8. Access/Security Considerations

Roads providing direct access to a recreational wastewater treatment facility should be constructed in a manner that minimizes accidents, and should include all-weather surfaces for immediate access at any time and season. Access roads must be clearly marked with gates or appropriate signs to discourage their use by the public. The facility should be enclosed by a chain-link fence to prevent children and animals from wandering into the facility area, and in general to deny access to the facility by the public. "Off-Limits to the Public" signs should be posted on the gate and fence. A telephone number should prominently displayed on all gates for emergencies.

## 7-9.  Comparison of Treatment Processes

This section presents an evaluation of the advantages/capabilities and disadvantages/limitations of small-scale wastewater treatment processes that are applicable to recreational areas. Table 7-1 presents the advantages/disadvantages of the conventional unit processes. See Chapter 5 for comparisons of natural systems and constructed wetlands. Table 7-2 lists operational characteristics of activated sludge processes currently available on the U.S. Army Engineer Waterways Experiment Station (WES) computing system.

**Table 7-1**
**Evaluation of Conventional Wastewater Treatment Processes**

| Treatment Processes | Application | Advantages and Capabilities | Disadvantages and Limitations |
|---|---|---|---|
| *a. Preliminary Treatment* | | | |
| (1) Screening | Waste streams containing large solids (wood, rags, etc.) | Prevents pump and pipe clogging<br><br>Reduces subsequent solids handling | Maintenance required to prevent screen plugging; ineffective for sticky solids |
| (2) Grit removal | Waste streams containing significant amounts of large, heavy, inorganic solids | Lowers maintenance costs and erosion<br><br>Reduces solids loading to other treatment units | Solids to be disposed of are sometimes offensive |
| (3) Equalization | Waste streams with variability | Dampens waste<br><br>Reduces chemical requirements<br><br>Dampens peak flows, reduces treatment plant size<br><br>Reduces corrosion and scaling | Needs large land areas<br><br>Possible septicity, requiring mixing and/or aeration equipment |
| (4) Temperature adjustment | Waste streams with temperatures | Provides the proper conditions for biological treatment | High initial equipment costs |
| (5) Nutrient | Nutrient deficient wastes | Optimizes biological treatment | High initial equipment costs |
| *b. Primary Treatment* | | | |
| (1) Sedimentation | Waste streams in settleable suspended solids | Reduces inorganic and organic solids loadings to subsequent biological units<br><br>By far the least expensive and most common method of solid-liquid separation<br><br>Suitable for treatment of a wide variety of wastes<br><br>Requires simplest equipment and operation<br><br>Demonstrated reliability as a treatment process | Possible septicity and odors<br><br>Adversely affected by variations in the nature of the waste<br><br>Moderately large area requirement |
| *c. Secondary Treatment* | | | |
| (1) Trickling filter | Biologically treatable organic wastes | Moderate quality (80-85% $BOD_5$ removal)<br><br>Moderate operating costs (lower than activated sludge and higher than oxidation pond)<br><br>Withstands shock loads better than other biological systems | High capital costs<br><br>Clogging of distributors or beds<br><br>Snail, mosquito, and insect problems |
| (2) Activated sludge (aeration and secondary sedimentation) | Biologically treatable organic wastes | Flexible, can adapt to minor pH, organic, and temperature changes<br><br>Small area required | High operating costs (skilled labor, electricity, etc.)<br><br>Generates solids requiring sludge disposal |

(Continued)

**Table 7-1 (Concluded)**

| Treatment Processes | Application | Advantages and Capabilities | Disadvantages and Limitations |
|---|---|---|---|
| *c. Secondary Treatment (continued)* | | | |
| (2) Activated sludge (aeration and secondary sedimentation) (Cont.) | | Degree of nitrification is controllable | Some process alternatives are sensitive to shock loads and metallic or other poisons |
| | | Relatively minor odor problems | |
| | | | Requires continuous air supply |
| (3) Aerated lagoon | Biologically treatable organic wastes | Flexible, can adapt to minor pH, organic, and temperature waste changes | Dispersed solids in effluent |
| | | Inexpensive construction | Affected by seasonal temperature variations |
| | | Minimum attention | Operating problems (ice, solids settlement, etc.) |
| | | Moderate effluent (50-75% BOD removal) | Moderate power costs |
| | | | Large area required |
| | | | No color reduction |
| (4) Oxidation Pond | Biologically treatable organic | Low construction cost | Large land area required |
| | | Nonskilled operation | Algae in effluent |
| | | Moderate treatment effectiveness (70-85% $BOD_5$ removal) | Possible septicity and odors |
| | | Removes some nutrients from wastewater | Weed growth, mosquito, and insects problems |
| (5) Anaerobic contact | Waste streams with high BOD and/or high temperature | Methane recovery | Heat required |
| | | Small area required | Effluent in reduced chemical form requires further treatment |
| | | Volatile solids destruction | Requires skilled operation |
| (6) Spray irrigation | Biologically treatable organic wastes | Inexpensive initial cost | |
| | | Minimum operator attention | |

Table 7-2
Operational Characteristics of Activated Sludge Processes

| Process Modification | Flow Model | Aeration System | $BOD_5$ Removal Efficiency | Application |
|---|---|---|---|---|
| Conventional | Plug-flow | Diffused-air, mechanical aerators | 85-95% | Low-strength domestic wastes, susceptible to shock labels. |
| Complete-mix | Complete-mix | Diffused-air, mechanical aerators | 85-95% | General application, resistant to shock loads, surface aerators. |
| Step-aeration | Plug-flow | Diffused-air | 85-95% | General application to wide range of wastes. |
| Modified-aeration | Plug-flow | Diffused-air | 60-75% | Intermediate degree of treatment where cell tissue in the effluent is not objectionable. |
| Contact-stabilization | Plug-flow | Diffused-air, mechanical aerators | 80-90% | Expansion of existing systems, package plants, flexible. |
| Extended-aeration | Complete-mix | Diffused-air, mechanical aerators | 75-85% | Small communities, package plants, flexible, surface aerators. |
| Kraus process | Plug-flow | Diffused-air | 85-95% | Low-nitrogen, high-strength wastes. |
| High-rate aeration | Complete-mix | Mechanical aerators | 75-90% | Use with turbine aerators to transfer oxygen and control floc size, general application. |
| Pure-oxygen systems | Complete-mix reactors in series | Mechanical aerators | 85-95% | General application, use where limited volume is available, use near economical source of oxygen, turbine or surface aerators. |

# Chapter 8
# Design References and Example

## 8-1. General

This chapter identifies wastewater treatment design manuals in which specific design criteria and examples can be found, and also lists design examples for small wastewater treatment plants not generally available in other references. These examples are not intended to be the only solution to the design of a small wastewater treatment facility in remote areas or for recreational purposes and are presented for information only. The purpose of this chapter (as well as Appendix E) is to provide the design engineer with the most up-to-date small-system wastewater treatment design information which has evolved since the latest Corps of Engineers engineering manual was published almost twenty years ago, and to emphasize earlier guidance which is still considered applicable. In identifying treatment methods applicable at recreational areas, design examples available in other military design manuals are referenced but not repeated herein. The treatment methods which follow are listed in the same general order as those methods discussed in Chapter 5.

## 8-2. Military Design Manuals

*a. General.* Design manuals which may be of assistance to the wastewater plant designer contain numerous examples outlining the approaches for collecting and analyzing the requisite information prior to the design phase of a project for both large and small wastewater facilities. A broad spectrum of design techniques, design considerations, and logical approaches are to be found in military publications.

*b. USACE manuals.*

(1) The most specific and current Corps of Engineers Manual, Army TM 5-814-3 (AFM 88-11, vol. 3), August 1988, provides general information, guidance and criteria for the design of domestic wastewater treatment facilities at permanent Army and Air Force installations located both in the United States and overseas. This manual discusses site selection; treatment requirements; basic design considerations; selection of treatment processes; small flow treatment systems such as septic tanks, waterless toilets, filtration/re-use systems, mound systems, Imhoff tanks, and package treatment plants; typical military wastewater treatment systems; plant layout; preliminary treatment; primary treatment; trickling filter plants; activated sludge plants; waste treatment ponds; advanced wastewater treatment; sludge handling; sludge treatment and disposal; disinfection (chlorination/dechlorination and ozone); flow measurement; sampling and process control; and considerations for both hot and cold climate operations.

(2) Design examples include: grit chamber; bar screen; proportional weir; Parshall flume; venturi flume; primary sedimentation; chemical precipitation; single-stage stone-media trickling filters and two-stage stone-media trickling filters; plastic media trickling filters; activated sludge (closed-loop reactor or oxidation ditch); microstrainer; multi-media filtration; activated carbon adsorption; phosphorous removal; nitrification-denitrification; anaerobic sludge digestion; aerobic sludge digestion; sludge pumping; gravity thickener; vacuum filtration; chlorination; and mound systems (pressure absorption fields for septic tanks). The manual includes pertinent design tables and accompanying figures.

## 8-3. National Small Flows Clearinghouse (NSFC) Publications

The NSFC was established in 1977 under the provisions of the Clean Water Act. In two decades of operation, the NSFC, funded by the U.S. Environmental Protection Agency and located at West Virginia University, Morgantown (1-800-624-8301; 304-293-3161), has become a national information source for small community wastewater treatment information. The engineering staff maintains extensive and comprehensive computer databases (bibliographic, facilities, regulatory, manufacturers and consultants, and state contacts). More than 250 books, brochures, case studies, database searches, and videotapes focusing on small community wastewater treatment issues are available; many are free, others no longer available from publishers are photocopied and priced on a cost-recovery basis. Of particular interest are treatment design manuals, septic tank pump pressure sewer systems (STEP) and design software for alternative sewers (pressure sewers, vacuum sewers, and small-diameter gravity systems). Selected references from NSFC were used to update this manual. For a complete list of publications, contact NSFC at the number listed above.

## 8-4. Wastewater Design Manuals and Texts

In the past decade, a number of comprehensive wastewater design texts and manuals have become available for designers and planners. The most recent publications, all of which are referenced in Appendix D, include:

- *Theory and Practice of Water and Wastewater Treatment*, Donald L. Droste, John Wiley & Sons, Inc., 1997 (Droste 1997).

- *Environmental Science and Engineering*, Second Edition, J. Glynn Henry and Gary W. Heinke, Prentice Hall, 1996 (Glynn 1996).

- *Unit Operations and Processes in Environmental Engineering*, Second Edition, Tom D. Reynolds and Paul A. Richards, PWS Publishing Company, Boston, Massachusetts, 1995 (Reynolds 1995).

- *Design of Municipal Wastewater Treatment Plants*, Volumes I and II, Water Environment Federation Manual of Practice No. 8 and ASCE Manual and Report on Engineering Practice No. 76, Book Press, Inc., 1991 (WEF MOP-8).

- *Wastewater Engineering Treatment, Disposal, and Reuse*, Metcalf and Eddy, Inc., Third Edition, revised by George Tchobanoglous and Franklin L. Burton, McGraw-Hill, Inc., 1991 (Metcalf & Eddy 1991).

## 8-5. U.S. Environmental Protection Agency (EPA)

EPA has kept current a series of wastewater-related design manuals, particularly in the following fields: wastewater treatment and disposal for small communities; alternative sewer systems; wetlands; land treatment; disinfection of wastewaters; biosolids/sludge handling, treatment and disposal. These publications, which are referenced in Appendix D, include:

*a. Small communities.*

"Wastewater Treatment and Disposal for Small Communities," EPA/625/R-92/005, September 1992.

*b. Alternative sewer systems.*

"Alternative Wastewater Collection Systems," EPA/625/1-91/024, October 1991.

*c. Wetlands.*

"Subsurface Flow Constructed Wetlands for Wastewater Treatment," EPA/832/R-93/008, July 1993.

"Water Quality Standards for Wetlands," EPA/400/S-90/011, July 1990.

"Constructed Wetlands and Aquatic Plant Systems for Municipal Wastewater Treatment," EPA/625/1-88/022, January 1988.

*d. Land treatment.*

"Land Treatment of Municipal Wastewater, Supplement on Rapid Infiltration and Overland Flow," EPA/1-81/013a, October 1984.

"Land Treatment of Municipal Wastewater," EPA/625/01-81/013, 1981.

*e. Disinfection.*

"Ultraviolet Radiation Technology Assessment," Office of Water, EPA-832-R-92-004, September 1992.

"Design Manual Municipal Wastewater Disinfection Source," EPA/625/1-86/021, October 1986.

*f. Sludge/Biosolids.*

"Surface Disposal of Sewage Sludges," Office of Water, EPA, May 1994 (USEPA-5).

"Land Application of Sewage Sludge," EPA/831-B-93-002b, December 1994.

"Biosolids Recycling," Office of Water, EPA, June 1994 (USEPA-4).

"Domestic Septage Regulatory Guidance," EPA/832-B-92-005, September 1993.

"A Plain English Guide to the EPA Part 503 Biosolids Rule," EPA/832/R-93/003.

"Control of Pathogens and Vector Attraction in Sewage Sludge," EPA/625/R-92/013, December 1992.

"Design Manual, Dewatering Municipal Wastewater Sludges," EPA/625/1-87/014, September 1987.

## 8-6. Wastewater Design Criteria and Example Matrices

A series of matrices has been developed to assist the design engineer in quickly locating pertinent references for the design of selected wastewater treatment and disposal systems. The following topics are addressed in Appendix D:

Conventional Wastewater Treatment-Preliminary Sedimentation, and Biological Processes (Table D-1).

Sludge Handling, Treatment and Disposal (Table D-2).

Small Wastewater Treatment Systems (Table D-3).

Natural Systems Wastewater Design Criteria and Examples (Table D-4).

Effluent Disinfection and Individual Treatment Processes Design Criteria and Examples (Table D-5).

## 8-7. Additional Design Examples

A number of small plant designs, particularly for selected secondary treatment processes, are not addressed in the references cited in the matrices in Appendix D. These include package plants, oxidation ditches (carousel type), stabilization ponds, sequencing batch reactors, and zero discharge. Hence, the following small plant design examples (Appendix E) are presented for these processes:

*a. Package plants (Extended-aeration activated sludge process).* The design for a 124 900 L/d (33,000 gal/d) extended-aeration package plant is shown as Example E-1. The package plant is a stand-alone unit, requires no pretreatment, but is equipped with an effluent disinfection process.

*b. Oxidation ditches (closed-loop reactors).* Oxidation ditches, or CLRs, are considered a secondary treatment process and generally require pretreatment. A design example for a "race-track" shaped CLR is provided in TM 5-814-3, Appendix C, Paragraph C-8. As CLRs may be built in place or prefabricated, a design for a "carousel" shaped facility to treat 379 000 L/d (100,000 gal/d) is provided as Example E-2.

*c. Stabilization ponds.* Stabilization ponds may be aerobic, facultative, or anaerobic, according to their oxygen profile. A 378 500 L/d (100 000 gal/d) stabilization pond with a primary clarifier and anaerobic digester of Imhoff Tank design with a secondary sand filter is given as Example E-3.

*d. Zero discharge by recycle/reuse (closed-loop reuse).* The closed-loop reuse principle is applicable to instances in which no liquid discharges from recreational treatment facilities are permitted or desired. After the system is initially filled and operational, any makeup wastewater from lavatories or drinking water fountains (estimated to represent about 6 percent of total water use) is allowed to evaporate from surface holding storage basins and the terminal holding pond or lagoon. Sludge is periodically removed from the surface holding storage basin. The design for a 37 900 L/d (10,000 gal/d) Zero Discharge Treatment Facility is shown as Example E-4.

*e. Sequencing batch reactors (SBR).* As stated in Chapter 5, the design of an SBR involves the same factors commonly used for the flow-through activated sludge system. If nitrification/dentrification and biological phosphorous removal are required, the SBR process must include pretreatment of the wastewater prior to the SBR reactor system. The design for a 379 000 L/d (100,000 gal/d) SBR treatment system is shown as Example E-5.

*f. Constructed wetlands.* The design of a 284 000 L/d (75,000 gal/d) aerobic non-aerated hyacinth constructed wetlands secondary treatment is shown as Example E-6.

# Chapter 9
# General Wastewater System Design Deficiencies

## 9-1. General

Design deficiencies of the several treatment processes mentioned in Chapters 5 and 8, where they occur, limit the performance of wastewater treatment plants. Elimination of deficiencies in the design phases of a project ensures that the final construction will incorporate the maximum number of operational conveniences for plant flexibility and process control. Deficiency reduction will permit more operable facilities which can be maintained at less cost and ensure that regulatory effluent standards are consistently met. Typical deficiencies in the design of various wastewater treatment systems are summarized in the following paragraphs. For more details consult USEPA-5.

## 9-2. Overall Considerations

a.   *Health/Safety/Security.*

(1)   Lack of hoists over larger pieces of equipment.

(2)   Lack of walkways around tanks, limiting operator access.

(3)   No provisions for moving equipment and supplies from one location to another.

(4)   Use of fixed louvers in buildings that cannot be shut during winter weather conditions.

(5)   Inadequate consideration of means to remove equipment for repair or replacement.

(6)   Inadequate communication capabilities between buildings and process areas.

(7)   Lack of all-weather roads to lift stations.

(8)   Inadequate clearance around equipment for maintenance functions.

(9)   No ladders or steps in manholes.

(10) Inadequate plant lighting.

(11) Stairways without non-skid surfaces.

(12) Inadequate hand railing and kick plates.

(13) Inadequate fencing and/or security gate around site.

(14) Use of air headers as guard railing at small package-plant type.

(15) Stairs inclined at too steep an angle.

(16) Guard railing not provided around ground-level tanks.

(17) Stairways provided with only one handrail.

(18) Valve handles located in unsafe areas.

(19) Dangerous chemicals not stored in separate areas.

(20) Hand rails and grating not secure.

(21) Stairs or steps not painted bright colors.

(22) Wet floors in some areas (pump rooms and pipe galleries) are slippery.

(23) Ladders in manholes and concrete tanks not secure.

(24) Hazardous areas not well defined.

(25) Permanent access platforms required for maintenance not provided.

(26) Interior building surfaces not painted with bright, easily-cleaned paints.

(27) Inadequate consideration of local weather conditions and their impact on the accessibility of a plant site.

(28) Failure to color code interior chemical feed lines.

(29) Inadequate consideration of spill prevention plan.

b. *Pumps.*

(1) Lack of spare pumps.

(2) Use of single-speed pumps where variable speed units are required.

(3) Pumps located above the normal water level, making them difficult to prime.

c. *HVAC.*

(1) Use of fixed louvers in buildings that cannot be shut during winter weather conditions.

(2) Lack of ventilation promotes corrosion of electrical components.

(3) Inadequate consideration of odor development and control.

(4) Inadequate consideration of ventilation requirements in confined spaces.

d. *Layout.*

(1) Inadequate flexibility to bypass units.

(2) Relative layout of process units and interconnecting piping not optimized.

(3) Layout of unit processes does not allow for future expansion of plant.

(4) Control panels not easily accessible (i.e., too high or placed in close quarters).

(5) Lack of flexibility to operate at low-flow start-up conditions.

(6) Electrical control panels located below ground where exposed to flooding.

(7) Inadequate consideration of potential freezing problems of plant components.

e. *Valves.*

(1) Inadequate valving for maximum flexibility and proper maintenance.

(2) Valves not operable from floor level.

(3) Lack of air bleed-off valves at high points in pump discharge lines.

(4) Lack of mud valves in tanks.

(5) Inadequate provisions for pressure relief around positive-displacement pumps.

(6) Inadequate consideration of the type of valve or gate used.

(7) Inadequate provisions for manual valve operation during emergency conditions.

f. *Piping.*

(1) Insufficient color coding of pipes and valves.

(2) Insufficient number and poor placement of high-pressure hose hydrants throughout plant.

(3) No provision for water tap at top of above-ground package units.

g. *Equipment.*

(1) Inadequate stand-by equipment.

(2) Stand-by generator either not provided or undersized to run all essential equipment during emergencies.

h. *Sampling.*

(1) Lack of sampling taps at pumping stations.

(2) Inadequate provisions for sampling of individual processes.

(3)  Lack of influent and/or composite sampler.

*i.  Sumps/Drains.*

(1)  Inadequate provisions for draining tanks and sumps.

(2)  Floor drain piping system undersized.

(3)  Drains from buildings discharge into basins with normally (or periodically) high-water levels, causing drains to back-up.

(4)  Lack of drains on chemical mix tanks.

(5)  Lack of sumps in dry wells.

*j.  Loadings.*

(1)  Foam sprays not concentrated in basin corners where foam buildup occurs.

(2)  Design based on average flow and $BOD_5$ and SS loadings with no recognition of peak conditions.

(3)  Lack of tank dewatering systems to permit rapid servicing of submerged equipment.

(4)  Excess oil from stationary units not contained.

(5)  Inadequate scum handling and disposal system.

*k.  Hydraulics.*

(1)  Undersized scum pits.

(2)  Inadequate consideration of pumping system design and/or fluid characteristics, resulting in pump cavitation.

(3)  Improper water pressure supplied to rota-meters.

*l.  Instrumentation.*

(1)  Lack of flow metering device on chemical feed lines.

(2)  Pressure gauges not located on inlet side of back-pressure relief valves, making it difficult to check and/or adjust the valve.

(3)  Lack of pressure gauges on plant pumps.

(4)  Inadequate number of flow meters.

*m. Electrical.*

(1) Absence of electrical outlets on top of treatment units.

(2) Electrical design does not have a power factor correction.

(3) Infrequent use of high-efficiency lighting sources.

(4) Motors oversized for future growth which never materializes, resulting in motors operating at less efficiency with lower power factors.

(5) Insufficient use of high-efficiency motors.

(6) Electrical quick-disconnect plugs not provided with submerged pumps to facilitate rapid replacement.

(7) Electric cut-off switches not locally mounted at individual pieces of equipment.

*n. Noise.*

(1) Inadequate noise abatement in various plant areas (i.e., blower, pump, and dewatering rooms, etc.).

(2) Inadequate consideration of noise control.

*o. Other.*

(1) Inadequate location of thrust blocks on pipe lines, particularly where couplings are involved or where automatic valves are located.

(2) Lack of cathodic protection for steel tanks.

(3) Lack of foam control system where required.

## 9-3. Conventional Design

*a. Preliminary treatment processes—general.*

(1) Inadequate consideration of pumping system design and/or fluid characteristics, resulting in pump cavitation.

(2) No provisions made to allow periodic cleaning of the influent wet well.

(3) Inadequate consideration of possible development of septic conditions in channels and splitter boxes.

(4) Lack of flexibility in disinfection systems to permit pre-chlorination for odor control or return sludge chlorination for control of bulking.

(5) High-water alarm system not provided.

b. *Primary treatment processes—general.*

(1) Lack of flexibility to operate at low-flow start-up conditions.

(2) Poor hydraulic and solids distribution among identical units operating in parallel.

(3) Undersized scum pits.

(4) Insufficient or inflexible sludge return and/or wasting pumping capacity.

(5) No mixing provided in scum tank to keep scum mixed during pumping.

(6) No positive method of removing scum from center well of clarifiers.

c. *Secondary treatment processes—general*

(1) Lack of walkways around tanks, limiting operator access.

(2) Inadequate consideration of scum removal from plant.

(3) Inadequate provisions for sampling of individual processes.

d. *Residuals hauling—general.*

(1) Use of single-speed pumps where variable-speed units are required.

(2) Undersized scum pits.

(3) Insufficient or inflexible sludge return and/or wasting pumping capacity.

(4) Lack of tank dewatering systems to permit rapid servicing of submerged equipment.

## 9-4. Preliminary Unit Processes

a. *Manual bar screen.*

(1) Lack of provision to remove floating material.

(2) Not locating grit removal and/or screening devices ahead of influent pumps to protect pumps from clogging or excessive abrasion.

(3) Inadequate consideration of proper disposal of coarse screenings and grit.

(4) Improper spacing of bars on bar screens.

(5) No provision for bypassing flow during maintenance.

(6) Improper velocity in bar screen chamber leading to grit deposition.

(7) Inadequate consideration of potential freezing.

*b. Mechanical bar screen.*

(1) Lack of provision to remove floating material.

(2) Not locating grit removal and/or screening devices ahead of influent pumps to protect pumps from clogging or excessive abrasion.

(3) Inadequate consideration of proper disposal of coarse screenings and grit.

(4) Improper spacing of bars on bar screens.

(5) No provision for bypassing flow during maintenance.

(6) Improper velocity in bar screen chamber leading to grit deposition.

(7) Inadequate consideration of potential freezing.

(8) Inadequate timing of mechanical rakes.

*c. Comminutor.*

(1) Not locating grit removal and/or screening devices ahead of influent pumps to protect pumps from clogging or excessive abrasion.

(2) Comminutor not located downstream of grit removal equipment, resulting in excessive cutting blade wear.

(3) No bar screen provided upstream for comminutor protection.

(4) No provision for bypassing flow during maintenance.

(5) No traps provided upstream of comminutor to still high velocity flows.

(6) Inadequate design permits grit deposits in control section of flow measurement device.

(7) Inadequate consideration of effect of waste material on mechanical reliability.

*d. Manually cleaned grit chamber.*

(1) Not locating grit removal and/or screening devices ahead of influent-pumps to protect pumps from clogging or excessive abrasion.

(2) Comminutor not located downstream of grit removal equipment, resulting in excessive cutting blade wear.

(3) Inadequate consideration of proper disposal of coarse screenings and grit.

(4) No provision for bypassing flow during maintenance.

(5) Inadequate velocity through process due to poor design.

(6) Improper flow-through velocity in grit chamber.

(7) Short-circuiting in grit chamber.

(8) Inadequate consideration of potential freezing.

e.   *Mechanically cleaned grit chamber.*

(1) Not locating grit removal and/or screening devices ahead of influent pumps to protect pumps from clogging or excessive abrasion.

(2) Comminutor not located downstream of grit removal equipment, resulting in excessive wear.

(3) Inadequate consideration of proper disposal of grit.

(4) No provision for bypassing flow during maintenance.

(5) Inadequate consideration of increased O&M and energy costs for grit collection process.

(6) Inadequate velocity through process due to poor flow control.

(7) Improper flow-through velocity in grit chamber.

(8) Short-circuiting in grit chamber.

f.   *Aerated grit chamber.*

(1) Not locating grit removal and/or screening devices ahead of influent pumps to protect pumps from clogging or excessive abrasion.

(2) Comminutor not located downstream of grit removal equipment, resulting in excessive wear.

(3) Inadequate consideration of proper disposal of grit.

(4) No provision for bypassing flow during maintenance.

(5) Inadequate consideration of increased O&M and energy costs for grit collection process.

(6) Improper flow-through velocity in grit chamber.

(7) Short-circuiting in grit chamber.

g.   *Grit pumps.*

(1)   Not locating grit removal and/or screening devices ahead of influent pumps to protect pumps from clogging or excessive abrasion.

(2)   Inadequate consideration of increased O&M and energy costs for grit collection process.

h.   *Influent flow measurement.*

(1)   Measurement control section not compatible with flow measurement device.

(2)   Inadequate design of downstream channel slope and geometry causes back-up in control section.

(3)   Inadequate design of obstructions downstream of control section induces inaccuracies in flow measurement.

(4)   Inadequate consideration of debris in wastewater in selection of float for flow measurement.

(5)   Flow meters located such that backwater elevation changes affect accuracy of meter.

(6)   Inadequate consideration of diurnal flow patterns in sizing of flow measurement equipment results in measurement equipment being inaccurate at the high and/or low flow ranges.

(7)   Inadequate approach channel length results in flow measurement inaccuracies.

(8)   Inadequate consideration of humidity in influent structure results in inaccuracies to flow sensor.

i.   *Raw wastewater pumping.*

(1)   Inadequate selection of the number, size, and type of pumps.

(2)   Inadequate provisions for removing scum from wet well.

(3)   No provisions for odor control in wet well of lift stations.

(4)   Not locating grit removal and/or screening devices ahead of influent pumps to protect pumps from clogging or excessive abrasion.

(5)   No provisions to periodically clean the wet well.

(6)   No bar screens provided for protection of mechanical components.

(7)   Inadequate design of pumping station results in frequent cycling of units, causing flow surges in downstream processes.

(8)   Lack of emergency overflow.

(9)   Improperly sized wet wells resulting in long detention times and odor problems, or too short detention time and cycling of pumps.

(10) Lack of spare air compressor for bubbler system.

(11) Inability to back-flush influent pumps for cleaning purposes.

(12) Corrosive and/or explosive gases close to electrical motors and equipment.

(13) Lack of proper ventilation at lift station.

*j.    In-line and side-line flow equalization.*

(1) Surface floating aerators do not allow basin to be dewatered.

(2) Inadequate or lack of facilities to flush solids and grease accumulations from the basin walls.

(3) Lack of facilities for withdrawing floating material and foam.

(4) Lack of emergency overflow.

(5) Lack of depth gauges provided on basins that operate at varying levels.

## 9-5. Primary Treatment Unit Process

*a.    Primary clarifier.*

(1) Improper length-to-width ratios.

(2) Inadequate clarifier sidewater depth.

(3) Design includes a common sludge removal pipe for two or more clarifiers, resulting in unequal sludge removal from the clarifiers.

(4) Effluent weir not uniformly level.

(5) Improper baffling resulting in short-circuiting causing inefficient solids removal.

(6) Septic conditions resulting from overloading or incorrect sludge removal.

(7) Inadequate consideration of impact of waste secondary sludge pumping on clarifier loading.

(8) Inadequate consideration of impact of various trickling filter recirculation rates and strategies on clarifier loadings.

(9) Inadequate consideration of clarifier inlet design.

(10) Inadequate sizing of torque requirement for sludge removal mechanism.

(11) Heavy wear on scrapers due to grit accumulations.

*b.    Primary sludge removal.*

(1)    Flushing and cleanout connections for sludge line not provided.

(2)    Primary sludge pumps located too far away from clarifiers.

(3)    Inadequate provisions for preventing frequent maintenance resulting from stringy or fibrous material in wastewater.

(4)    Inadequate provisions for chain, flight, and sprocket repair and replacement.

(5)    Design includes a common sludge removal pipe for two or more clarifiers, resulting in unequal sludge removal from the clarifiers.

(6)    Inadequate provisions for sampling of raw sludge.

(7)    Operator is not provided with the capability to observe sludge while pumping.

(8)    Inadequate flexibility in sludge pumping system.

(9)    No provisions for measuring sludge flow.

(10)    Inadequate consideration of character of sludge in sizing and layout of sludge lines.

(11)    Flushing and cleanout connections for sludge line not provided.

(12)    Primary sludge pumps located too far away from clarifiers.

(13)    Improper sizing of increments on time clock results in pumping of unnecessarily thin sludge.

*c.    Scum removal.*

(1)    Inadequate provisions for preventing frequent maintenance resulting from stringy material in wastewater.

(2)    Improper placement of scum removal equipment hinders clarifier performance.

(3)    Scum is recycled through the plant and not removed from the system.

(4)    Improper selection of scum pumping facilities results in excessive O&M.

## 9-6.  Secondary Treatment Unit Processes

*a.    Secondary clarification.*

(1)    No provision for addition of chemicals to improve settling characteristics.

(2)    Improper type of sludge removal mechanism selected.

(3) Improper clarifier sidewater depth.

(4) Inadequate access to weirs for sampling and maintenance.

(5) Inadequate consideration of impact and control of in-plant side streams.

(6) Overflow rate (OFR) of clarifiers too high to meet effluent suspended solids limitations.

(7) No provisions for flow division boxes.

(8) Short-circuiting in clarifiers.

(9) Improper weir placement (i.e., proper weir length but closely placed troughs create high, localized upward velocities within clarifier).

(10) Improper length-to-width ratio.

(11) Inadequate or no provisions for scum removal from secondary clarifiers.

(12) Long scum lines frequently become clogged.

(13) Scum will not flow from scum tanks once supernatant is pumped out.

(14) Sludge lines periodically clog, and no back-flush facilities are provided.

(15) Inability to conveniently dewater scum puts.

(16) Inadequate consideration of freezing problems and effect of cold temperatures on efficiency of biological treatment.

(17) Sludge collection equipment inadequately sized.

b.  *Trickling filter.*

(1) General.

(a) Improper design and installation of rotary distribution arms cause clogging and rotation problems.

(b) Side wall not high enough to prevent splashing or aerosol drifting.

(c) Lack of flexibility to flood the filter.

(d) Poor ventilation of filter under drains which may cause odor problems and/or inadequate oxygen for sustainable biological growth.

(e) Clogging of distributor orifices caused by inadequate preliminary or primary treatment.

(f) Inflexibility in flow patterns and/or recirculation strategy.

(g) Inadequate consideration of overspray on filter walls resulting in fly problems.

(h) Inadequate sizing of filter units to meet a more stringent effluent limitations requirement.

(i) Insufficient flow, particularly during low flow conditions, to rotate the distribution arms.

(j) Recirculation of secondary clarifier effluent through filters causes high flows through the clarifier, resulting in clarifier solids carry-over.

(k) No provision for flushing underdrain system.

(l) Inadequate or too frequent recirculating flow to filter causes media plugging.

(2) Rock media.

(a) Improper selection of media without good weathering properties.

(b) Inadequate air circulation provided during periods of high flows.

(c) Inadequate or too frequent recirculating flow to filter causes media plugging.

(d) Ice buildup on filter media.

(3) Plastic media.

(a) Inadequate air circulation provided during periods of high flows.

(b) Inadequate or too frequent flow to filter causes media plugging.

(c) Ice buildup on filter media.

(4) Distribution of wastewater.

(a) Improper design and installation of distribution arms cause clogging and rotation problems.

(b) Lack of flexibility to flood the filter.

(c) Clogging of distributor orifices caused by inadequate preliminary or primary treatment.

(d) Inadequate flow-dosing equipment.

(e) Insufficient flow, particularly during low-flow conditions, to rotate the distribution arm.

(f) Inadequate freeze protection.

(5) Flow recirculation.

(a) Inability to adjust, measure, and control recirculation rate.

(b)   Lack of proper recirculation pumping capacity.

(c)   Recirculation of secondary clarifier effluent causes high flows through the clarifier, resulting in clarifier solids carryover.

(d)   Inadequate consideration of effects of recirculation through primary clarifiers on clarifier loadings.

c.   *Rotating biological contactors.*

(1)   Bearings located below grade make RBCs susceptible to flooding.

(2)   Buildings not insulated and facility heat losses in winter cause wastewater temperature to drop, thereby reducing biological activity.

(3)   Primary clarifiers not provided, causing settling and plugging of media.

(4)   Excessive detention time in pre-RBC channels promote the development of septic conditions.

(5)   Side streams not accounted for in design of RBC units.

(6)   Inadequate screening of raw wastes causes plugging of RBC media.

(7)   Inefficient tank design causes dead spots and solids deposition in RBC tank.

(8)   Improper design of overflow baffles between stages causes solids deposition.

d.   *Air activated sludge.*

(1)   General.

(a)   Lack of flexibility to operate in different modes (i.e., complete mix, plug flow, contact-stabilization, etc.).

(b)   Aerator spacing not adequately considered.

(c)   Inadequate foam control throughout activated sludge basin lengths.

(d)   Inadequate mixing prevents solids deposition and uniform suspended solids and dissolved oxygen concentrations throughout the basin.

(e)   Inadequate preliminary screening of raw wastes causes plugging of aerators and return/waste sludge pumping system.

(f)   Inflexible design does not permit isolation of reactors and changes in flow schemes for maintenance purposes and/or to adjust for changes in wastewater characteristics.

(g)   Inadequate consideration of impact and control of in-plant side streams.

(h)   Inadequate provisions for bypassing aeration basin for repair.

(i)   Improper sidewater depth and baffling cause splashing problems in basin.

(j)   Inability to control and measure mixed-liquor flow distribution to multiple secondary clarifiers.

(k)   Inadequate consideration of impact of changing aeration basin levels on aerator performance.

(l)   Multi-compartmental basins do not have reinforced inner walls; therefore, individual tanks cannot be dewatered.

(m)   Inability to drain foam spray system results in freezing problems.

(n)   Supports for air drop pipes cannot be seen when aeration basin is full, making it difficult to reinstall the drop pipes.

(o)   Inadequate aeration capacity.

(p)   Lack of splash shields in front of effluent gates.

(q)   Inadequate consideration of freezing problems and effect of cold temperatures on efficiency of biological treatment.

(2)   Diffusers.

(a)   Inadequate or no air cleaners provided on blowers results in plugging of diffusers.

(b)   No provisions for removing air diffuser drop pipes from aeration tanks.

(c)   Air valves not graduated to allow even distribution of air flow to diffusers.

(3)   Fixed mechanical aerators.

(a)   Improper placement of gear box drains causes oil to drain into aeration basin.

(b)   Amp meters not provided at motor control center so operators cannot tell if proper amperage is being drawn.

(c)   No time delay relays provided to limit stress shock to aerator gears when shifting from high speed to low speed.

(4)   Floating aerators.

(a)   Floating aerators located too close to wall or pontoons not aligned properly, causing pontoons to strike the basin wall when starting up.

(b)   Improper placement of gear box drains causes oil to drain into aeration basin.

(c) Amp meters not provided at motor control center so operator cannot tell if proper amperage is being drawn.

(5) Blowers.

(a) Inadequate or no air cleaners provided on blowers.

(b) Blower silencers not provided.

(6) Dissolved oxygen control and measurement.

(a) Inability to adequately measure and adjust air flow rates to control dissolved oxygen levels and energy consumption.

(b) Improper design of dissolved oxygen measuring instrumentation does not allow easy removal of equipment for routine inspection and maintenance.

(7) Return sludge pumping.

(a) Inadequate provisions for sampling and observation of return and waste-activated sludge.

(b) Improper selection of valves for sludge lines.

(c) Improper return sludge flow splitting.

(d) Inadequate sludge recycle/waste capacity.

(e) Inadequate sludge flow measurement for small plants using air lift pumps.

(f) Inability to adjust, measure, and control return/waste sludge flows due to lack of instrumentation.

(g) Inability to change placement of return sludge in aeration basin.

(h) Scum accumulation in flow splitter boxes.

(8) Waste sludge pumping.

(a) No separate waste sludge pumps.

(b) Inadequate provisions for sampling and observation of return and waste activated sludge.

(c) Improper selection of valves for sludge lines.

(d) Inadequate waste sludge pipe sizing for "slip-stream" wasting.

(e) Inadequate sludge recycle/waste capacity.

(f) Inadequate sludge flow measurement for small plants using air lift pumps.

## 9-7. Sludge Dewatering

(a) Improper placement of control panels in spray/splash areas hampers clean-up and results in high corrosion rates.

(b) Inadequate consideration of storage of dewatered sludge during inclement weather.

(c) Inadequate consideration of potential plugging problems in sludge piping.

(d) Inadequate consideration of corrosive nature of materials to be handled.

(e) Tank drain lines are located 50-76 mm (2-3 in.) off the bottom of tanks, making it difficult to dewater the basins completely.

(f) Clogging problems in lime piping.

(g) Inadequate provisions for vibration control in sludge piping design.

(h) Sludge pumping and dewatering areas not properly ventilated.

(i) Inadequate provisions for lifting equipment for repairs.

## 9-8. Non-Conventional Plants

*a. Package plants.*

(1) Lack of spare pumps.

(2) Lack of walkways around tanks, limiting operator access.

(3) Inadequate flexibility to bypass units.

(4) Inadequate consideration of means to remove equipment for repair or replacement.

(5) Inadequate consideration of scum removal from plant.

(6) Inadequate laboratory facilities for process control.

(7) Inadequate standby equipment.

(8) Inadequate provisions for draining tanks and sumps.

(9) Inadequate scum handling and disposal system.

(10) Foam sprays not concentrated in basin corners where foam buildup occurs.

(11) No provision for water tap at top of above-ground package units.

(12) Use of constant speed pumps where variable-speed units are required.

(13) No provisions made to allow periodic cleaning of the influent wet well.

(14) Samplers frequently clog.

(15) Lack of mud valves in tanks.

(16) Lack of a foam control system.

(17) Absence of electrical outlets on top of treatment units.

(18) Lack of auxiliary power.

(19) Inadequate hand railing and kick plates.

(20) Stairways provided with only one handrail.

(21) Inadequate consideration of noise control.

(22) Inadequate consideration of potential freezing problems of plant components.

b. *Ponds and lagoons.*

(1) Inadequate valving for maximum flexibility and proper maintenance.

(2) Inadequate process flexibility.

(3) Inadequate consideration of access requirements for large equipment (cranes, trucks, etc.) required for maintenance.

(4) Individual flow measurement not provided for each piece of parallel units.

(5) Inadequate consideration of possible development of septic conditions in channels and splitter boxes.

(6) Inadequate provisions for manual valves for emergency conditions.

(7) Inability of process to meet effluent requirements in winter.

(8) Inadequate (or lack of) liner to meet state requirements, and to prevent groundwater pollution.

(9) Single-point entry into pond overloads pond in feed zone.

(10) Lack of multiple cells for operating flexibility.

(11) Anaerobic conditions due to organic overloading.

(12) No drains provided in ponds or lagoons.

(13) Water level gauges not provided.

(14) Improper vertical depth between lagoon bottom and groundwater table.

(15) No groundwater monitoring wells provided.

## 9-9. Land Application

a.   *Overland flow slope design.*

(1)   Improper slope construction.

(2)   Inadequate detention time on slope to achieve desired level of treatment.

(3)   Inappropriate location of land treatment plots.

(4)   Inadequate soil depth for suitable land treatment.

(5)   Inadequate site loading for optimum treatment.

(6)   Inadequate shaping of drainage channels for efficient system operation.

(7)   Improper selection of maintenance equipment to minimize soil compaction.

(8)   Inadequate location of service roads.

b.   *Cover crop.*

(1)   Inadequate detention time on slope to achieve desired level of treatment.

(2)   Improper selection of maintenance equipment to minimize soil compaction.

c.   *Hydraulic application.*

(1)   Improper slope construction.

(2)   Inadequate soil depth for suitable land treatment.

(3)   Inadequate site loading for optimum treatment.

(4)   Inadequate knowledge of subsurface drainage alternatives to alleviate drainage problems.

(5)   Improper selection of maintenance equipment to minimize soil compaction.

d.   *Soil depth.*

(1)   Inadequate soil depth for suitable land treatment.

(2)   Inadequate consideration given to soil type and the interaction of soil with sodium in the wastewater.

e.  *Infiltration beds.*

(1)  Improper slope construction.

(2)  Inadequate soil depth for suitable land treatment.

(3)  Inadequate construction of lagoons for maintenance and sludge removal.

(4)  Inadequate site loading for optimum treatment.

f.  *Odor control.*

(1)  Improper slope construction.

(2)  Inadequate consideration for needs of pre-chlorination or pre-aeration.

g.  *Center pivot sprinkler.*

(1)  Spray nozzles plug due to solids in wastewater.

(2)  Inadequate protection of equipment for freezing conditions.

(3)  Inadequate pumping facilities for control of sedimentation in piping.

(4)  Inadequate facilities provided for flushing of lateral lines.

(5)  Inadequate selection of protective coatings to minimize corrosion.

(6)  Plastic laterals installed above-ground break because of cold weather.

h.  *Traveling gun sprinkler.*

(1)  Spray nozzles plug due to solids in wastewater.

(2)  Inadequate pumping facilities for control of sedimentation in piping.

(3)  Inadequate facilities provided for flushing of lateral lines.

(4)  Inadequate sprinkle head design to minimize aerosolization.

(5)  Plastic laterals installed above-ground break because of cold weather.

## 9-10.  Sludge Drying and Disposal

a.  *Sludge drying beds.*

(1)  Inadequate drainage system.

(2)  No provisions for cake removal from sand bed.

(3)   Inadequate provisions for proper sludge distribution.

(4)   Inadequate layout of underdrains.

(5)   Improper location of sand bed allows inflow of surface drainage.

(6)   Inadequate consideration of potential flooding of sand bed.

(7)   Improper sand gradation.

(8)   Walls dividing sludge drying beds are made of untreated wood and warp rapidly.

(9)   Inadequate freeze protection.

(10)  Inadequate consideration of local climate on dewatering rate and size requirements for sand beds.

b.   *Sludge disposal.*

(1)   Inadequate consideration of sludge concentration/transportation tradeoffs.

(2)   Inadequate consideration of equipment utility in all-weather conditions.

(3)   Lack of vector control.

(4)   Inadequate consideration of nutrients and public health hazards (metals, bacteria) transport in soil/groundwater.

(5)   Inadequate buffer zone at disposal site.

(6)   Lack of odor control/prevention.

(7)   Sludge loading delayed due to lack of truck or container capacity.

c.   *Composting.*

(1)   Inadequate space for sludge staging and preparation.

(2)   Inadequate sludge storage during maintenance periods.

(3)   Inadequate consideration of feed solids concentration.

(4)   Inadequate consideration of fresh air supply and overall ventilation requirements.

(5)   Inadequate provisions for reliable auxiliary fuel source.

(6)   Inadequate consideration of ultimate residue disposal.

(7)   Inadequate odor control.

## 9-11.  Sewer Collection Systems

(a)  Failure to specify proper construction materials on sewer lines, e.g., cast iron pipe across creek or when elevated on piers.

(b)  Failure to provide vented covers on manholes located on high ground.

(c)  Failure to provide tight lids on low-ground manhole covers.

## 9-12.  Lift Stations

(a)  Failure to locate lift stations on protected side of streams to reduce possible flooding.

(b)  Failure to provide access ladders to all wet wells.

(c)  Failure to slope bottoms of all wet wells.

(d)  Failure to provide solid covers for wet wells and means of securing same.

(e)  Failure to fence lift stations where locations require security.

(f)  Failure to provide standby power for lift stations.

(g)  Failure to vent all wet wells.

(h)  Inadequate consideration of lift station valving.

# Chapter 10
# Sludge Disposal

## 10-1. General

*a. Overview.* The disposal of sludges and septage generated at treatment facilities is an important consideration in the selection and design of such facilities. This chapter provides an overview of current sludge disposal requirements and restrictions.

*b. Current standards.* The current EPA Standards for the Use or Disposal of Sewage Sludge became effective March 22, 1993 (40 CFR Part 503). The new standards attempt to unravel problems that heretofore have caused a number of inconsistencies in the use and disposal of wastewater-generated sludges. These standards address three principal sludge issues: land-applied, distributed, or marketed sludge; disposal at dedicated sites or in sludge-only landfills (monofills); and incineration in sludge-only incinerators. The new standards also affect septage disposal. Generally, burial in landfills, either monofills or with municipal wastes, has been found to be the most cost-effective method of disposal.

## 10-2. Definitions

"Sludge" is defined as the residual material removed from wastewater treatment facilities. A new term, "biosolids," suggests the beneficial usage of sludge. The definition of "biosolids" is now accepted as those primarily organic solid products produced by wastewater treatment processes that can be beneficially recycled. Other definitions of interest follow. (Note: the definitions apply specifically to the 1993 rule and may differ from definitions previously provided.)

- A Class I sludge management facility is any publicly owned treatment works, including federally owned treatment works.

- Domestic septage is either liquid or solid material removed from a septic tank, cesspool, portable toilet, marine sanitation device, or similar treatment works that receives only domestic wastewater. Domestic septage does not include liquid or solid material removed from a septic tank, cesspool, or similar treatment works that receives either commercial wastewater or industrial wastewater and does not include grease removal from grease traps.

- Domestic wastewater is waste and wastewater from humans or operations that is discharged to, or otherwise enters, a treatment works.

- A person is an individual, association, partnership, corporation, municipality, state or Federal agency, or an agent or employee thereof.

- Treatment of sludge is its preparation for final use or disposal including, but not limited to, thickening, stabilization, and dewatering of sludge. Sludge treatment does not relate to storage of wastewater sludge.

- Wetlands means those areas inundated or saturated by surface water or groundwater at a frequency and duration to support, and that under normal circumstances do support, a prevalence of vegetation typically adapted for life in saturated solid conditions. Wetlands include swamps, marshes, bogs, and similar areas.

## 10-3. Management Standards

*a. Rule 40 CFR 503.* The 40 CFR 503 rule includes standards that apply to generators, processors, beneficial users, or disposers of sludge or septage wastes. Briefly, the rule establishes two new national standards. First, based on risk assessment, it sets standards for 10 heavy metals, pathogens (mainly disease-causing viruses, bacteria, and parasites), and emissions from incinerators (not under consideration at remote recreational or isolated areas). Second, there is a standard for managing septage and sewage sludge and for their disposal. Prescribed management practices were designed to limit human and ecological exposure to contaminants and then to ensure that the sludge produced is used on the land or is properly disposed of in a manner that protects both human health and the local environment.

*b. General practices.* General management practices include pollutant monitoring, pathogen reduction, vector attraction reduction, site restrictions, protection of threatened or endangered species, and record keeping.

*a. Specific practices—land application.* Specific management practices for land application include the use of an agronomic rate of application based on the needs of site-specific crops; application in ways that prevent runoff to waters of the U.S. including a buffer zone of 10 m; labeling or instructions for those who purchase sludge-derived products for individual use; and site restrictions.

*b. Specific practices—surface disposal.* Specific management practices for surface disposal methods include more stringent site restrictions, including those on grazing animals, crops, and human contact; certification or monitoring to ensure no contamination of groundwater; air monitoring for methane gas; and runoff collection requirements.

## 10-4. Toxic Metal Regulations

The 40 CFR 503 rule contains limits for 10 metal pollutants for land application: arsenic, cadmium, chromium, copper, lead, mercury, molybdenum, nickel, selenium, and zinc. The rule also specifies limits for 3 toxic metals for surface disposal: arsenic, chromium, and nickel.

See paragraph 10-10 for surface disposal pollutant limits.

## 10-5. Effect of Land Application

The new rule contains two regulatory strategies for land application depending on the quality of the sludge in question. Sludges that are shown or proven to be of "exceptionally high quality" become exempt from further regulatory controls and can be used as freely as other soil amendments or fertilizer would be. Sludges of good quality that do not meet the "exceptionally high quality" standards can also be used on land if certain management practices are observed.

## 10-6. Pathogen and Vector Attraction Reduction

Pathogen and vector attraction reduction requirements are major changes from previous federal sludge regulations and include two classes of pathogen reduction: Class A and Class B. Class A is a Process to Further Reduce Pathogens (PFRP) standard; Class B is a Process to Significantly Reduce Pathogens (PSRP) standard (see paragraph 10-11).

## 10-7. Exclusions

The new rule does not apply to sludges co-landfilled with solid waste, sludges co-incinerated with solid waste, grit and screenings, and drinking water treatment sludges. Sewage scum is not excluded and is discussed in paragraph 10-14; any scum collected in wastewater clarifier operations falls within the definition of sludge.

## 10-8. Land Application Pollutant Limits

Maximum allowable metals concentrations established in the 1993 rule for land application of sludge are shown in Table 10-1. Specifically:

- Sludge cannot be applied if the pollutant concentration exceeds the maximum allowable concentrations.

- If sludge is applied to land, either the cumulative loading rate or pollutant concentrations must not be exceeded.

- If sludge is applied to a lawn or home garden, the pollutant concentrations must not be exceeded.

- If sludge is sold or given away in a bag or container, the pollutant concentration or the application rate must not be exceeded.

**Table 10-1**
**Regulatory Limits For Toxic Metals**

| Pollutant | Pollutant Ceiling Concentrations [mg/kg] | Cumulative Pollutant Loading Rates [kg/ha (lb/acre)] | Pollutant Concentrations (mg/kg) | Annual Pollutant Loading Rates [kg/ha - yr(lb/acre-yr)] |
|---|---|---|---|---|
| Arsenic | 75 | 41 (37) | 41 | 2 (1.8) |
| Cadmium | 85 | 39 (35) | 39 | 1.9 (1.7) |
| Chromium | 3000 | 3000 (2700) | 1200 | 150 (135) |
| Copper | 4300 | 1500 (1350) | 1500 | 75 (40.5) |
| Lead | 840 | 300 (270) | 300 | 15 (13.5) |
| Mercury | 57 | 17 (15) | 17 | 0.85 (0.77) |
| Molybdenum | 75 | 18 (16) | 18 | 0.90 (0.81) |
| Nickel | 420 | 420 (378) | 420 | 21 (19) |
| Selenium | 100 | 100 (90) | 36 | 5 (4.5) |
| Zinc | 7500 | 2800 (2520) | 2800 | 140 (126) |

## 10-9. Land Application Management Practices

Sludge shall not be land applied if:

- It is likely to adversely affect threatened or endangered species;

- The land is flooded, frozen, or snow-covered so that with resulting runoff the sludge enters a wetland or other waters;

- It is 10 m (33 ft) or less from surface waters; or

- It exceeds the agronomic rate, unless it is applied to a permitted reclamation site.

For sludges sold or given away in bags or other containers, an information sheet or label shall be provided to the user. The label or information sheet shall contain the following information:

- Name and address of the generator.

- The annual application rate at which the sludge may be applied.

- A statement that the application rate shall not be exceeded.

## 10-10.  Surface Disposal Pollutant Limits

The general requirements for surface disposal are:

- Disposal is prohibited in wetlands or unstable areas.

- Disposal is prohibited when the site is within 60 m (200 ft) of a geologic fault unless a special permit is obtained.

- Closure and post-closure plans are required six (6) months prior to site closure.

- There are no pollutant limits for surface disposal sites with liners and leachate collection systems.

- There are pollutant limits for arsenic, chromium, and nickel for surface disposal sites without liners and leachate collection.  The limit depends on the distance from property boundaries as indicated in Table 10-2.

For surface disposal, there are certain required management practices as well as frequency of monitoring requirements.

## 10-11.  Pathogens and Vector Attraction Reduction

There are differing requirements for Class A and Class B pathogens and vector attraction reductions as follows:

*a. Class A pathogen reduction.*  All options require pathogen reduction to indicate that the sludge has either: <1 000 MPN fecal coliforms per gram total solids, or <3 MPN Salmonella per four grams of total solids; and one of the following six alternatives:  control time and temperature, raise the sludge pH, reduce enteric viruses and helminth ova (low pathogen sludge), reduce enteric viruses and helminth ova (normal sludge), process to further reduce pathogens treatment (see paragraph 10-12), and process to further reduce pathogens equivalent treatment (see paragraph 10-12).

**Table 10-2**
**Maximum Pollutant Concentrations for Surface Disposal**

| Site to Property Line Distance (m) | Pollutant Concentrations[1] (mg/kg) | | |
|---|---|---|---|
| | Arsenic | Chromium | Nickel |
| >150 | 73 | 600 | 420 |
| 125 to < 150 | 62 | 450 | 420 |
| 100 to < 125 | 53 | 360 | 390 |
| 75 to < 100 | 46 | 300 | 320 |
| 50 to < 75 | 39 | 260 | 270 |
| 25 to < 50 | 34 | 220 | 240 |
| 0 to < 25 | 30 | 200 | 210 |

[1] Applies only to sites without liners and leachate collection systems.

*b. Class B pathogen reduction.*

(1) There are three options for Class B pathogen reduction: <2 000 000 MPN coliforms per gram total solids (geometric mean of seven samples); PSRP (Process to Significantly Reduce Pathogens treatment (see paragraph 10-12); and PSRP equivalent treatment (see paragraph 10-12).

(2) Five site restrictions also apply:

• Food crops—no harvesting after sludge application for 14 to 38 months depending upon type of crop grown and how sludge is applied.

• Feed crops—no harvesting for 30 days after sludge application.

• Pasture—no animal grazing for 30 days after sludge application.

• Turf—no harvesting for one year after sludge application

• Public access—restricted access for 30 days (after 30 days for low exposure areas; one year for high exposure areas).

(3) There are twelve methods of vector attraction reduction for land application, surface disposal, and septage (see Table 10-3).

## 10-12. Pathogen Treatment Processes

Composting of sewage sludges is covered in the pathogen treatment processes.

*a. Process to significantly reduce pathogens.* When composting is practiced using either the in-vessel composting method or the windrow composting method, the temperature of the sewage sludge is to be raised to 40°C (104°F) or higher and remain at 40°C (104°F) or higher for five days. For four hours during the five days, the temperature in the pile must exceed 55°C (131°F).

Table 10-3
Vector Attraction Reduction

| Method | Practice | | |
|---|---|---|---|
| | Land Application | Surface Disposal | Septage |
| 38 % volatile solids (VS) reduction | • | • | |
| Bench test for low VS anaerobic sludge | • | • | |
| Bench test for low VS aerobic sludge | • | • | |
| Specific oxygen uptake rate (SOUR) < 1.5 mg O$_2$/hr/g | • | • | |
| 14 days temp > 40° C (>104° F) avg. Temp > 45° C (>113° F) | • | • | |
| pH > 12 for 2 hours and pH > 11.5 for an additional 22 hours | • | • | |
| 75% dry solids (DS) (no primary treatment) | • | • | |
| 90% DS | • | • | |
| Subsurface injection | * | • | • |
| Incorporation | * | • | • |
| Daily Cover | | • | † |
| pH > 12 for 30 min. | | | • |

*b. Process to further reduce pathogens.* When composting using either the within-vessel composting method or the static aerated pile composting method, the temperature of the sewage sludge is maintained at 55°C (131°F) or higher for three days. Using the windrow composting method, the temperature of the sewage sludge is to be maintained at 55°C (131°F) for 15 days or longer. During the period when the compost is maintained at 55°C (131°F) or higher, there shall be a minimum of five (5) turnings of the windrow.

## 10-13. Septage Applied to Agricultural Land, Forests, or Reclamation Sites

Part 503 imposes separate requirements for domestic septage applied to agricultural land, forests, or a reclamation site. If domestic septage is applied to public contact sites or home lawns or gardens, the same requirements must be met as for bulk biosolids which are land applied (general requirements, pollutant limits, pathogen and vector attraction reduction requirements, management practices, frequency monitoring requirements, as well as record keeping and reporting requirements).

## 10-14. Wastewater Scum

*a. Components.* Scum is a minor component of wastewater solids collection and consists of all materials that float to the liquid surface of unit processes. In the primary sedimentation or clarification units, scum generally includes little biological foam or skimmings. The secondary sedimentation or clarification process produces almost entirely biological scum and foam and is generally minimal in volume.

*b. Disposal.*

(1) Scum disposal is regulated by the 1991 EPA 40 CFR 258 (Criteria for Municipal Solid Waste Landfills) as well as the 1993 EPA 40 CFR 503 (Standards for the Use and Disposal of Sewage Sludge). Part 503 defines sewage sludge as, but not limited to, "...domestic septage; scum and solids removed in primary, secondary or advanced wastewater treatment processes; and a material derived from sewage

sludge." The description is apt, as scum has many of the characteristics of wastewater solids/residues. Scum may be disposed with sewage sludge.

(2) Under the 1991 EPA 40 CFR 258 rule, sewage sludge and scum may be disposed of in municipal solid waste landfills and are jointly defined as "...any solid, semi-solid, or liquid waste generated from a municipal, commercial, or industrial wastewater treatment plan...."

(3) Under the 1993 EPA 40 CFR 503 rule, land application and surface disposal is regulated and specifically refers to composting practices. Any scum included in surface disposal or land application must meet certain pollutant limits, Class A or Class B pathogen requirements, or vector reduction requirements and some additional management requirements.

## 10-15. Composting Methods

*a. Static pile process.* Schematics for an aerated static pile, a conventional windrow, and an aerated windrow composting process are shown in Figure 10-1. The extended-aerated static pile process involves mixing the dewatered sludge with a bulking agent followed by active composting in specially constructed piles as shown in Figure 10-2.

*b. Windrow process.* The conventional windrow process shown in Figure 10-1 involves initial mixing of dewatered sludge with a bulking agent such as finished compost and supplemented with an external amendment followed by the formation of long windrows. The aerated windrow process is similar to the conventional windrow process with one exception, i.e., a system for induced aeration is provided in addition to aeration by turning with a mobile composter.

*c. Compositing considerations.* Important considerations in static pile composting include wood chip usage, initial mixing, pile construction, composting period, process control, drying, screening, and curing. Considerations for windrow composting include windrow formation, composting period, drying, and dust generation.

## 10-16. Composting Additives/Amendments/Bulking Agents

Addition of carbonaceous materials in the composting process may resolve other unrelated disposal problems. Where tree cutting, leaf collection, and storm debris cleanup present a disposal problem, some of this material can be used in the composting process as carbonaceous additives. Alternately, many readily available materials such as sawdust, bark or wood chips, shredded newsprint, rice or peanut hulls, and corncobs have been successfully utilized as carbonaceous additives. Leaves, straw, and shredded tires have also been successfully used. If the material is biodegradable, thus helping to promote biological activity, it is called an amendment. Bulking agents are primarily used to provide structural support and maintain air space in the windrows and piles.

## 10-17. Equipment

Equipment at composting operations varies considerably. Turning piles in windrow composting is accomplished by commercial rototillers, front-end loaders, or specially designed turning machines such as Cobey Composters.

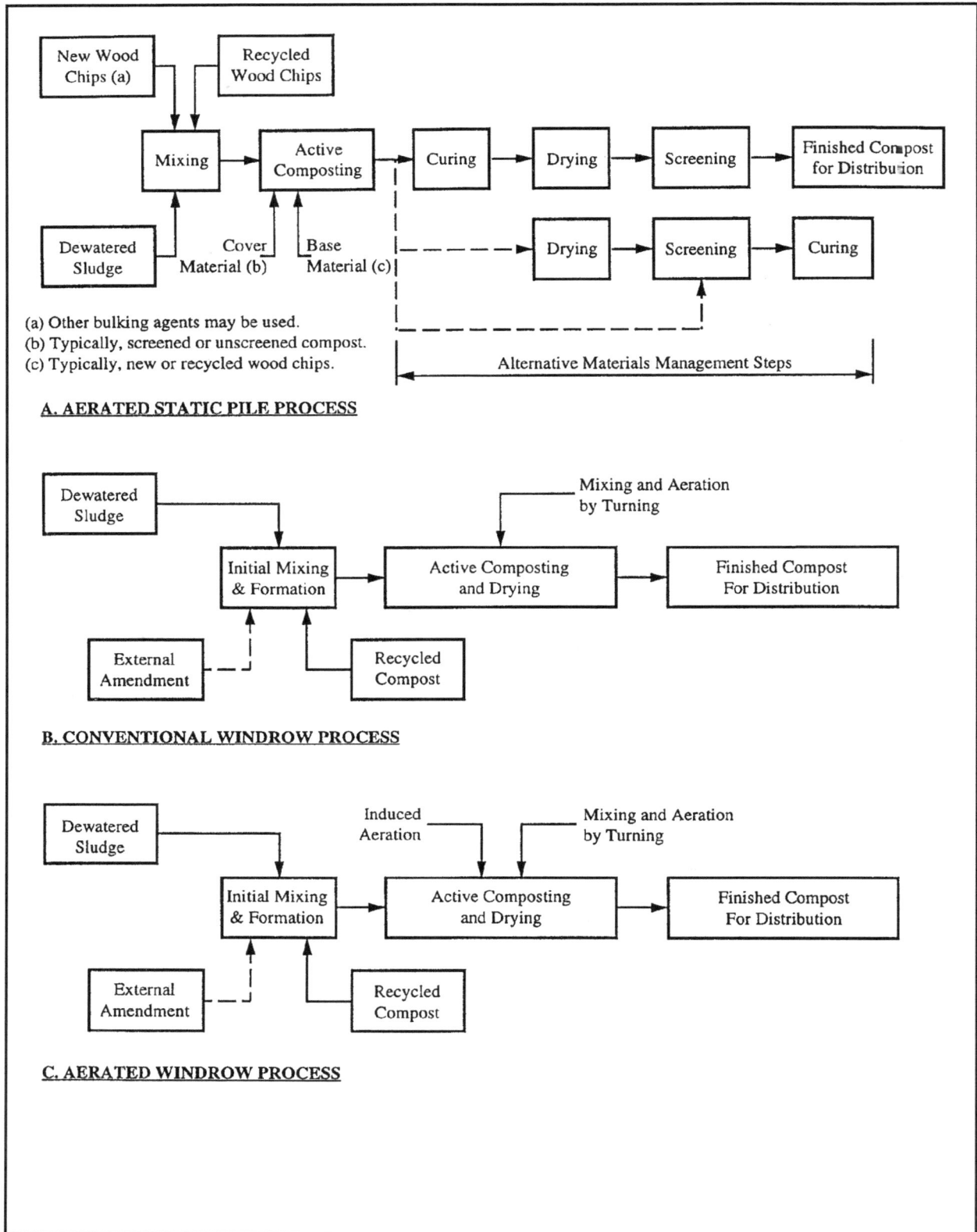

```
New Wood          Recycled
Chips (a)         Wood Chips

         Mixing  →  Active    →  Curing →  Drying →  Screening →  Finished Compost
                    Composting                                    for Distribution

Dewatered       Cover      Base                       Drying →  Screening →  Curing
Sludge          Material   Material (c)
                (b)
```

(a) Other bulking agents may be used.
(b) Typically, screened or unscreened compost.
(c) Typically, new or recycled wood chips.

Alternative Materials Management Steps

**A. AERATED STATIC PILE PROCESS**

```
Dewatered                              Mixing and Aeration
Sludge                                 by Turning

              Initial Mixing  →  Active Composting  →  Finished Compost
              & Formation         and Drying             For Distribution

External          Recycled
Amendment         Compost
```

**B. CONVENTIONAL WINDROW PROCESS**

```
Dewatered               Induced        Mixing and Aeration
Sludge                  Aeration       by Turning

              Initial Mixing  →  Active Composting  →  Finished Compost
              & Formation         and Drying             For Distribution

External          Recycled
Amendment         Compost
```

**C. AERATED WINDROW PROCESS**

Figure 10-1.  Sludge compositing process schematics

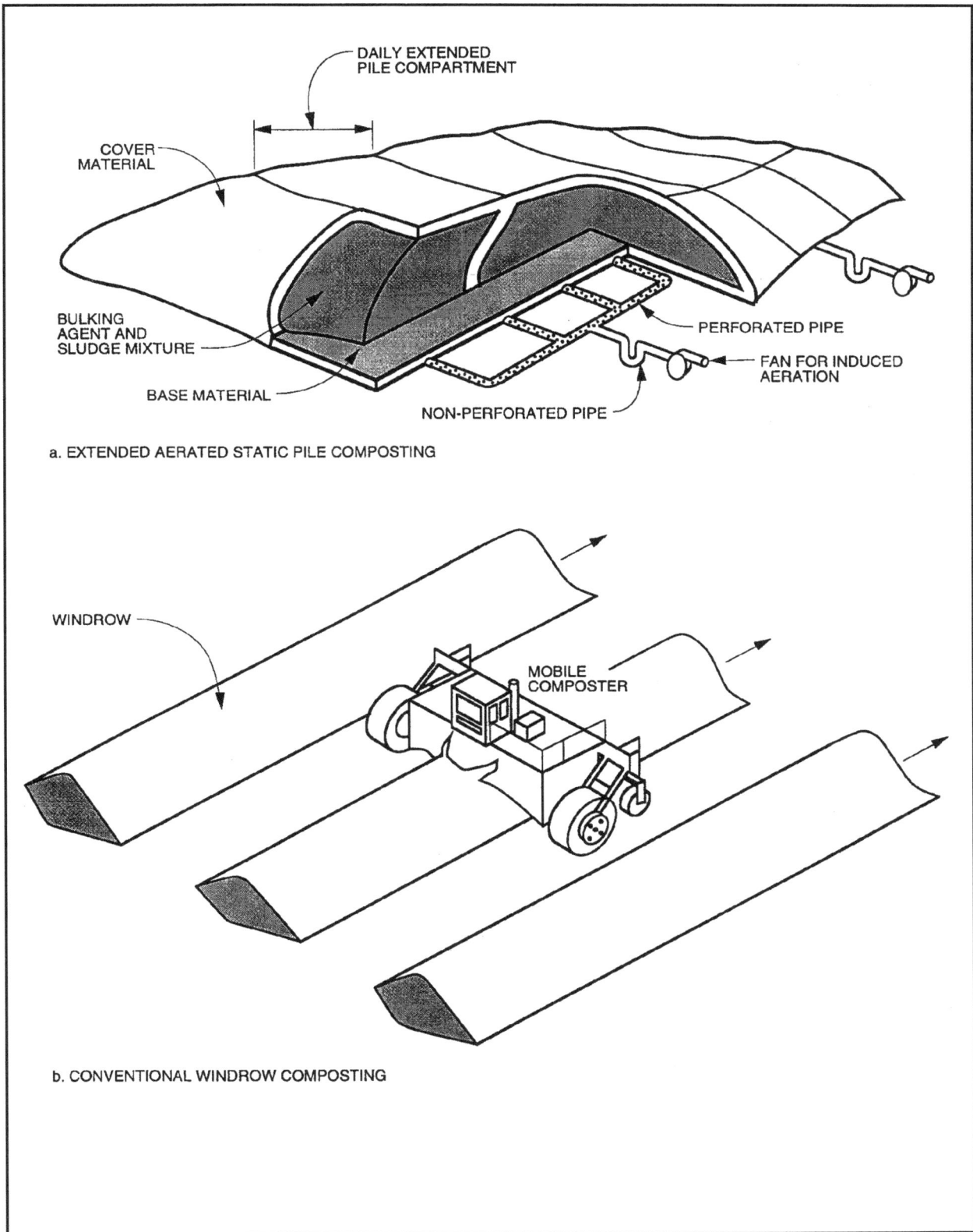

a. EXTENDED AERATED STATIC PILE COMPOSTING

b. CONVENTIONAL WINDROW COMPOSTING

Figure 10-2.   Sludge compositing methods

## 10-18. Guidance

The best U.S. Army document relating to sludge handling and disposal at military installations is TM 5-814-3. Other pertinent sources for sludge/biosolids are found in EPA/332/R-93/003, EPA/625/R-92/013, and EPA/831/B-93/002b.

# Appendix A
# References

## A-1. Required Publications

**ER 1110-1-261**
Quality Assurance of Laboratory Testing Procedures

**ER 1110-1-8100**
Laboratory Investigations and Testing

**EM 200-1-3**
Requirements for the Preparation of Sampling and Analysis Plans

**EM 385-1-1**
Safety and Health Requirements Manual

**EM 1110-2-504**
Land Treatment Systems Operation and Maintenance

**ETL 1110-3-442**
Ultraviolet Disinfection at Army Wastewater Treatment Facilities

**TM 5-814-1**
Sanitary an Industrial Wastewater Collection—Gravity Sewers and Appurtenances, March 1985.

**TM 5-814-2**
Sanitary and Industrial Wastewater Collection—Pumping Stations and Force Mains, March 1985.

**TM 5-814-3**
Domestic Wastewater Treatment, August 1988.

**TM 5-814-8**
Evaluation Criteria Guide for Water Pollution Prevention, Control, and Abatement Programs, 23 April 1987.

## A-2. Related Publications

**AOAC 1982**
Statistical Manual of the Association of Official Analytical Chemists, "Statistical Techniques for Collaborative Tests, Planning and Analysis of Results of Collaborative Tests," 1982. Available from Association of Official Analytical Chemists, 400 Nath Frederick Ave., Suite 500, Gaithersburg, MD 20877-2417 (FAX: (301) 924-7089).

**AR 200-1**
Environmental Quality, Environmental Protection and Enhancement.

**Burs 1994**
Burs, Bennette D., and Minnis, Mary Margaret, "Onsite Wastewater Treatment Systems," Hogarth House Ltd., Madison, WI, 1994. Available from Hogarth House, Ltd., 1715 Madison St., Madison, WI 53711.

**Clark 1971**
Clark, Viessman, and Hammer, "Water Supply and Pollution Control," 2nd edition, International Textbook Co., Scranton, PA, 1971. The International Textbook Company has been taken over by International Correspondence Schools, 925 Oak St., Scranton, PA 18505. Address inquiries there.

**Converse 1990**
Converse, James C., and Tyler, E. Jerry, "Mound Systems, Design Module," University of Wisconsin - Madison, January 1990. Available from The Kurt F. Wendt Library, University of Wisconsin, 2.5 North Randall St., Madison, WI 53706.

**Corbitt 1990**
Corbitt Robert A., "Standard Handbook of Environmental Engineering," McGraw-Hill Publishing Co., 1990. Available from The McGraw-Hill Companies, 1221 Avenue of the Americas, New York, NY, 10020.

**Droste 1997**
Droste, Donald L., "Theory and Practice of Water and Wastewater Treatment," John Wiley & Sons, Inc., 1997. Available from John Wiley and Sons, Inc., 605 Third Avenue, New York, NY 10158-0012, or 1 Wiley Drive, Somerset, NJ 08875-1272.

**Ehlers 1965**
Ehlers, V. M. And Steel, E. W., "Municipal and Rural Sanitation," McGraw-Hill, New York, 1965. Available from the McGraw-Hill Companies, 1221 Avenue of the Americas, New York, NY 10020.

**EPA-625-R-92/005**
Wastewater Treatment and Disposal for Small Communities, September 1992. Available from National Technical Information Service, U.S. Department of Commerce, 5285 Port Royal Road, Springfield, VA 22161 (FAX: (703) 321-8547), or USEPA National Center for Environmental Publications and Information, P.O. Box 42419, Cincinnati, OH 45242-2419 (FAX: (513) 891-6685).

**EPA/625/1-91/024**
Alternative Wastewater Collection Systems, October 1991. Available from National Technical Information Service, U.S. Department of Commerce, 5285 Port Royal Road, Springfield, VA 22161 (FAX: (703) 321-8547), or USEPA National Center for Environmental Publications and Information, P.O. Box 42419, Cincinnati, OH 45242-2419 (FAX: (513) 891-6685).

**EPA/825/1-80/012**
Design Manual: Onsite Wastewater Treatment and Disposal Systems, October 1980. Available from National Technical Information Service, U.S. Department of Commerce, 5285 Port Royal Road, Springfield, VA 22161 (FAX: (703) 321-8547), or USEPA National Center for Environmental Publications and Information, P.O. Box 42419, Cincinnati, OH 45242-2419 (FAX: (513) 891-6685).

**EPA/625/R-92/010**

Small Community Water and Wastewater Treatment, September 1992. Available from National Technical Information Service, U.S. Department of Commerce, 5285 Port Royal Road, Springfield, VA 22161 (FAX: (703) 321-8547), or USEPA National Center for Environmental Publications and Information, P.O. Box 42419, Cincinnati, OH 45242-2419 (FAX: (513) 891-6685).

**EPA/625/1-84/013a**

Process Design Manual for Land Treatment of Municipal Wastewater: Supplement on Rapid Infiltration and Overland Flow, 1984. Available from National Technical Information Service, U.S. Department of Commerce, 5285 Port Royal Road, Springfield, VA 22161 (FAX: (703) 321-8547), or USEPA National Center for Environmental Publications and Information, P.O. Box 42419, Cincinnati, OH 45242-2419 (FAX: (513) 891-6685).

**EPA/1-81/013**

Process Design Manual for Land Treatment of Municipal Wastewater, 1981. Available from National Technical Information Service, U.S. Department of Commerce, 5285 Port Royal Road, Springfield, VA 22161 (FAX: (703) 321-8547), or USEPA National Center for Environmental Publications and Information, P.O. Box 42419, Cincinnati, OH 45242-2419 (FAX: (513) 891-6685).

**EPA/625/1-88/022**

Constructed Wetlands and Aquatic Plant Systems for Municipal Wastewater, January 1988. Available from National Technical Information Service, U.S. Department of Commerce, 5285 Port Royal Road, Springfield, VA 22161 (FAX: (703) 321-8547), or USEPA National Center for Environmental Publications and Information, P.O. Box 42419, Cincinnati, OH 45242-2419 (FAX: (513) 891-6685).

**EPA/832/R-93/008**

Subsurface Flow Constructed Wetlands for Wastewater Treatment, 1993. Available from National Technical Information Service, U.S. Department of Commerce, 5285 Port Royal Road, Springfield, VA 22161 (FAX: (703) 321-8547), or USEPA National Center for Environmental Publications and Information, P.O. Box 42419, Cincinnati, OH 45242-2419 (FAX: (513) 891-6685).

**EPA/400/S-90/013a**

Water Quality Standards for Wetlands. Available from National Technical Information Service, U.S. Department of Commerce, 5285 Port Royal Road, Springfield, VA 22161 (FAX: (703) 321-8547), or USEPA National Center for Environmental Publications and Information, P.O. Box 42419, Cincinnati, OH 45242-2419 (FAX: (513) 891-6685).

**EPA/1-81/013**

Land Treatment of Municipal Wastewater, Supplement on Rapid Infiltration and Overland Flow, October 1984. Available from National Technical Information Service, U.S. Department of Commerce, 5285 Port Royal Road, Springfield, VA 22161 (FAX: (703) 321-8547), or USEPA National Center for Environmental Publications and Information, P.O. Box 42419, Cincinnati, OH 45242-2419 (FAX: (513) 891-6685).

**EPA/625/01-81/013**

Land Treatment of Municipal Wastewater, 1981. Available from National Technical Information Service, U.S. Department of Commerce, 5285 Port Royal Road, Springfield, VA 22161 (FAX: (703) 321-8547), or USEPA National Center for Environmental Publications and Information, P.O. Box 42419, Cincinnati, OH 45242-2419 (FAX: (513) 891-6685).

**EPA-832-R-92-004**
Ultraviolet Radiation Technology Assessment, Office of Water, September 1992. Available from National Technical Information Service, U.S. Department of Commerce, 5285 Port Royal Road, Springfield, VA 22161 (FAX: (703) 321-8547), or USEPA National Center for Environmental Publications and Information, P.O. Box 42419, Cincinnati, OH 45242-2419 (FAX: (513) 891-6685).

**EPA/625/1-86/021**
Design Manual—Municipal Wastewater Disinfection Source, October 1986. Available from National Technical Information Service, U.S. Department of Commerce, 5285 Port Royal Road, Springfield, VA 22161 (FAX: (703) 321-8547), or USEPA National Center for Environmental Publications and Information, P.O. Box 42419, Cincinnati, OH 45242-2419 (FAX: (513) 891-6685).

**EPA/832-B-92-005**
Domestic Septage Regulatory Guidance, September 1993. Available from National Technical Information Service, U.S. Department of Commerce, 5285 Port Royal Road, Springfield, VA 22161 (FAX: (703) 321-8547), or USEPA National Center for Environmental Publications and Information, P.O. Box 42419, Cincinnati, OH 45242-2419 (FAX: (513) 891-6685).

**EPA/332/R-93/003**
A Plain English Guide to EPA Part 503 Biosolids Rule, September 1994. Available from National Technical Information Service, U.S. Department of Commerce, 5285 Port Royal Road, Springfield, VA 22161 (FAX: (703) 321-8547), or USEPA National Center for Environmental Publications and Information, P.O. Box 42419, Cincinnati, OH 45242-2419 (FAX: (513) 891-6685).

**EPA/625/R-92/013**
Control of Pathogens and Vector Attraction in Sewage Sludge, December 1992. Available from National Technical Information Service, U.S. Department of Commerce, 5285 Port Royal Road, Springfield, VA 22161 (FAX: (703) 321-8547), or USEPA National Center for Environmental Publications and Information, P.O. Box 42419, Cincinnati, OH 45242-2419 (FAX: (513) 891-6685).

**EPA/625/1-87/014**
Design Manual, Dewatering Municipal Wastewater Sludges. Available from National Technical Information Service, U.S. Department of Commerce, 5285 Port Royal Road, Springfield, VA 22161 (FAX: (703) 321-8547), or USEPA National Center for Environmental Publications and Information, P.O. Box 42419, Cincinnati, OH 45242-2419 (FAX: (513) 891-6685).

**EPA/831/B-93/002b**
Land Application of Sewage Sludge, December 1994. Available from National Technical Information Service, U.S. Department of Commerce, 5285 Port Royal Road, Springfield, VA 22161 (FAX: (703) 321-8547), or USEPA National Center for Environmental Publications and Information, P.O. Box 42419, Cincinnati, OH 45242-2419 (FAX: (513) 891-6685).

**Francingues 1976**
Francingues, N. R., Jr., and Green, A. J., Jr., "Water Usage and Wastewater Characterization at a Corps of Engineers Recreation Area," Miscellaneous U.S. Paper Y-76-1, January 1976, U.S. Army Engineer Waterways Experiment Station, Vicksburg, MS. Available from USAEWES, 3909 Halls Ferry Road, Vicksburg, MS 39180-6199.

**Glynn 1996**
Henry, J. Glynn, and Heinke, Gary W., "Environmental Science and Engineering," 2nd Edition, Prentice Hall, 1996. Available from Prentice Hall, 200 Old Tappan Road, NJ 07675.

**Hammer 1989**
Hammer, Donald A., "Constructed Wetlands for Wastewater Treatment: Municipal, Industrial, and Agricultural," Lewis Publishers, 1989. Available from Lewis Publishers, 2000 Corporate Boulevard, NW, Boca Raton, FL 33431.

**Harrison 1972**
Harrison, J., "Disposal of Vault Wastes, Lake Ouachita and Lake Greeson," MiscellaneoU.S. Paper Y-72-1, March 1972, U.S. Army Engineer Waterways Experiment Station, CE, Vicksburg, Mississippi. Available from USAEWES, 3909 Halls Ferry Road, Vicksburg, MI 39180-6199.

**Hydromantis 1992**
CAPDET-PC, "A Computer-Assisted Program for the Design of Wastewater Treatment Facilities," Version 2.07, 1992. Available from Hydromantis, Inc., 1685 Main Street West, Suite 302, Hamilton, Ontario, Canada, L85IG5.

**Kaplan 1989**
Kaplan, O. Benjamin, "Onsite Wastewater Disposal," Lewis Publishers, 1989. Available from Lewis Publishers, 2000 Corporate Boulevard, NW, Boca Raton, FL 33431.

**Kaplan 1991**
Kaplan, O. Benjamin, "Septic Systems Handbook," Lewis Publishers, 1991. Available from Lewis Publishers, 2000 Corporate Boulevard, NW, Boca Raton, FL 33431.

**Matherly 1975**
Matherly, J., et al., "Water Usage and Wastewater Characteristics, Shelbyville, Illinois," 1975, U.S. Army Construction Engineering Research Laboratory, Champaign, IL. Available from U.S. Army Construction Engineering Laboratory Library, P.O. Box 9005, Champaign, IL 61826.

**Metcalf & Eddy 1972**
Metcalf and Eddy, Inc., "Wastewater Engineering; Collection, Treatment, and Disposal," McGraw-Hill, New York, 1972. Available from the McGraw-Hill Companies, 1221 Avenue of the Americas, New York, NY 10020, or 860 Taylor Station Road, Blacklick, OH 43004-0545, or 13311 Monterey Avenue, Blue Ridge Summit, PA 17294-0850.

**Metcalf & Eddy 1991**
"Wastewater Engineering Treatment, Disposal and Reuse," Metcalf and Eddy, Inc., Third Edition, revised by George TchobanogloU.S. and Franklin L. Burton, McGraw-Hill, Inc., 1991. Available from The McGraw-Hill Companies, 1221 Avenue of the Americas, New York, NY 10020, or 860 Taylor Station Road, Blacklick, OH 43004-0545, or 13311 Monterey Avenue, Blue Ridge Summit, PA 17294-0850.

**Middleton USACE**
Middleton, Braxier, and Bolden, "Basic Sewage Characteristics at a Corps of Engineers Recreation Area," U.S. Army Engineer District, St. Louis, CE, St. Louis, MO. Available from U.S. Army Corps of Engineers, St. Louis District, 1222 Spruce Street, St. Louis, MO 63103-2825.

**NSFC-1**
National Small Flows Clearinghouse, Additives Information Package, WWPCGN66. Available from National Small Flows Clearinghouse, P.O. Box 6064, West Virginia University, Morgantown, WV 26506-6064 (1-800-624-8301).

**NSFC-2**
National Small Flows Clearinghouse, "Constructed Wetlands and Aquatic Plant Systems for Municipal Wastewater." Available from National Small Flows Clearinghouse, P.O. Box 6064, West Virginia University, Morgantown, WV 26506-6064 (1-800-624-8301).

**NSFC-3**
National Small Flows Clearinghouse, "Constructed Wetlands Wastewater Treatment Systems for Small Users Including Individual Residences," May 1993. Available from National Small Flows Clearinghouse, P.O. Box 6064, West Virginia University, Morgantown, WV 26506-6064 (1-800-624-8301).

**NSFC-4**
National Small Flows Clearinghouse, "Graywater Systems From the State Regulations," October 1996. Available from National Small Flows Clearinghouse, P.O. Box 6064, West Virginia University, Morgantown, WV 26506-6064 (1-800-624-8301).

**NSFC-5**
National Small Flows Clearinghouse, "No-Flow Toilets From the State Regulations," March 1995. Available from National Small Flows Clearinghouse, P.O. Box 6064, West Virginia University, Morgantown, WV 26506-6064 (1-800-624-8301).

**NSFC-6**
National Small Flows Clearinghouse, "Percolation Tests From the State Regulations," November 1996. Available from National Small Flows Clearinghouse, P.O. Box 6064, West Virginia University, Morgantown, WV 26506-6064 (1-800-624-8301).

**NSFC-7**
National Small Flows Clearinghouse, "Application Rates & Sizing of Fields From the State Regulations," November 1996. Available from National Small Flows Clearinghouse, P.O. Box 6064, West Virginia University, Morgantown, WV 26506-6064 (1-800-624-8301).

**NSFC-8**
National Small Flows Clearinghouse, "State Design Criteria for Wastewater Treatment Systems," September 1990. Available from National Small Flows Clearinghouse, P.O. Box 6064, West Virginia University, Morgantown, WV 26506-6064 (1-800-624-8301).

**OSHA 1996**
29, CFR Parts 1903, 1904, and 1910, Department of Labor, Occupational Safety and Health Administration (OSHA), 1996. Available from U.S. Government Printing Office, Main Bookstore, 710 N. Capitol Street, NW, Washington, DC 20401 (1-888-293-6498).

**OSU 1992**
"Mound Systems Pressure Distribution of Wastewater," Ohio State University, 1992. Available from The Science and Engineering Library, Ohio State University, 175 West 18th Avenue, Columbus, OH 43210.

**Penn Bureau of Resources**
Pennsylvania Bureau of Resources Programming, "Engineering Report, Sanitary Fixture Requirements for Pennsylvania Park Facilities," 1972, Philadelphia, PA.

**Reynolds 1995**
Reynolds, Tom D., and Richards, Paul A., "Unit Operations and Processes in Environmental Engineering," Second Edition, PWS Publishing Company, Boston, MA, 1995. Available from PWS Publishing Company, A Division of Wadsworth, Inc., 20 Park Plaza, Boston, MA 02116.

**Robin and Green**
Robin and Green, "Development of Onshore Treatment Systems for Sewage from Watercraft Waste Retention System," U.S. EPA, Cincinnati, Ohio. Available from National Technical Information Service, U.S. Department of Commerce, 5285 Port Royal Road, Springfield, VA 22161 (FAX: (703) 321-8547), or USEPA National Center for Environmental Publications and Information, P.O. Box 42419, Cincinnati, OH 45242-2419 (FAX: (513) 891-6685).

**Simmons 1972**
Simmons, G. M. and Phen, R. L., "Forest Service Sanitary Waste Facilities," 1972, California Institute of Technology, Pasadena, California. Available from the Robert A Milliken Manual Library 1-32, California Institute of Technology, 1200 E. California Boulevard, Pasadena, CA 91125.

**Smith 1973**
Smith, S. and Wilson, J. "Truck Wastes," "Water and Wastes Engineering," March 1973, pp 48-57.

**Taylor 1963**
Taylor, D. W., "Soil Mechanics," John Wiley and Sons, Inc., New York, 1963. Available from John Wiley and Sons, Inc., 1 Wiley Drive, Somerset, NJ 08875-1272.

**Teraghi 1960**
Teraghi, K. and Peck, R. B., "Soil Mechanics in Engineering Practice," John Wiley and Sons, Inc., New York, 1960. Available from John Wiley and Sons, Inc., 605 Third Avenue, New York, NY 10158-0012, or 1 Wiley Drive, Somerset, NJ 08875-1272.

**USACE 1986**
PC-ECONPACK, "Automated Economic Analysis Package," prepared by the U.S. Army Corps of Engineers, Huntsville Division, June 1986. Available from U.S. Army Engineer and Support Center, PO Box 1600, Huntsville, AL 35807-4301.

**USACERL 1984**
"Appropriate Technology for Treating Wastewater at Remote Sites on Army Installations: Preliminary Findings," U.S. Army Engineer Construction Engineering Research Laboratories, Champaign, Illinois, Technical Report N-160, April, 1984. Available from U.S. Army Construction Engineering Laboratory Library, P.O. Box 9005, Champaign, IL 61826.

**USACERL 1993**
"An Evaluation of Reed Bed Technology to De-water Army Wastewater Treatment Plant Sludge," U.S. Army Engineer Construction Engineering Research Laboratories, USACERL Technical Report EP-93/09, September 1993. Available from U.S. Army Construction Engineering Laboratory Library, PO Box 9005, Champaign, IL 61826.

**USAEWES**

Francigues, N.R., Jr,. "Design of Water and Sewer Systems and Treatment Methods for Public Use Areas—Information Summary," U.S. Army Engineer Waterways Experiment Station, Vicksburg, MS. Available from USAEWES, 3909 Halls Ferry Road, Vicksburg, MS 39180-6199.

**USDA-1**

Briar Cook, "Guidelines for the Selection of a Toilet Facility," U.S. Department of Agriculture, 9123 1204, April, 1991. Available from Publications Office, U.S. Department of Agriculture, Room 506-A, 1400 Independence Avenue, SW, Washington, DC 20250.

**USDA-2**

Brenda L., "Land Remote Waste Management," U.S. Department of Agriculture, 9523 1202, May 1995. Available from Publications Office, U.S. Department of Agriculture, Room 506-A, 1400 Independence Avenue, SW, Washington, DC 20250.

**USDA-3**

"Updated Vault Toilet Concepts," U.S. Department of Agriculture. Available from Publications Office, U.S. Department of Agriculture, Room 506-A, 1400 Independence Avenue, Washington, DC 20250.

**USDA-4**

Briar Cook, "In-Depth Design and Maintenance Manual for Vault Toilets," U.S. Department of Agriculture, 9123-1601, July, 1991. Available from Publications Office, U.S. Department of Agriculture, Room 506-A, 1400 Independence Avenue, Washington, DC 20250.

**USDA-5**

Brenda L. "Composting Toilet Systems, Planning, Design, and Maintenance, Land, U.S. Department of Agriculture, 9523 1803, July 1995. Available from Publications Office, U.S. Department of Agriculture, Room 506-A, 1400 Independence Avenue, Washington, DC 20250.

**USDHEW 1967**

U.S. Department of Health, Education and Welfare, Public Health Service, "Manual of Individual Water Supply Systems," Publication No. 24 1967, Washington, DC Available from Department of Health and Human Services, ATTN: PHS-Public Affairs, 200 Independence Avenue, SW, Washington, DC 20201.

**USDOI 1958**

U.S. Department of Interior, National Park Service, "National Park Building Constructor Handbook," 1958, Washington, DC. Available from National Park Service Technical Information Center, ATTN: DSC-MS, P.O. Box 25287, Denver, CO 80225-0287 (FAX: (303) 969-2557).

**USEPA-1**

Glueckert, A. J. and Seigh, P. A., "Onshore Treatment System for Sewage from Watercraft Retention Systems," U.S. Environmental Protection Agency, Cincinnati, Ohio. Available from National Technical Information Service, U.S. Department of Commerce, 5285 Port Royal Road, Springfield, VA 22161 (FAX: (703) 321-8547), or USEPA National Center for Environmental Publications and Information, PO Box 42419, Cincinnati, OH 45242-2419 (FAX: (513) 891-6685).

**USEPA-2**

"Handbook for Analytical Quality Control in Water and Wastewater Laboratories," June 1972, Analytical Quality Control Laboratory, Cincinnati, OH 45268. Available from National Technical Information Service, U.S. Department of Commerce, 5285 Port Royal Road, Springfield, VA 22161 (FAX: (703) 321-8547), or USEPA National Center for Environmental Publications and Information, P.O. Box 42419, Cincinnati, OH 45242-2419 (FAX: (513) 891-6685).

**USEPA-3**

"Surface Disposal of Sewage Sludges," Office of Water, EPA, May 1994. Available from National Technical Information Service, U.S. Department of Commerce, 5285 Port Royal Road, Springfield, VA 22161 (FAX: (703) 321-8547), or USEPA National Center for Environmental Publications and Information, P.O. Box 42419, Cincinnati, OH 45242-2419 (FAX: (513) 891-6685).

**USEPA-4**

"Biosolids Recycling, Office of Water," EPA, June 1994. Available from National Technical Information Service, U.S. Department of Commerce, 5285 Port Royal Road, Springfield, VA 22161 (FAX: (703) 321-8547), or USEPA National Center for Environmental Publications and Information, P.O. Box 42419, Cincinnati, OH 45242-2419 (FAX: (513) 891-6685).

**USEPA-5**

"Identification and Correction of Typical Design Deficiencies at Municipal Wastewater Treatment Facilities," prepared for the Office of Water, U.S. Environmental Protection Agency, by Roy F. Weston, Inc. Available from National Technical Information Service, U.S. Department of Commerce, 5285 Port Royal Road, Springfield, VA 22161 (FAX: (703) 321-8547), or USEPA National Center for Environmental Publications and Information, P.O. Box 42419, Cincinnati, OH 45242-2419 (FAX: (513) 891-6685).

# Appendix B
# States with Regulations/Requirements Applicable to Small-Scale Wastewater Treatment Facilities

---

**Table B-1**
**State Regulatory Contacts**

---

*Alabama*
Department of Public Health
Division of Community Environmental Protection
434 Monroe St.
Montgomery, AL 36130-1701
(334) 613-5373

*Alaska*
Department of Environmental Conservation
Domestic Wastewater Program
410 Willoughby Ave., Suite 105
Juneau, AK 99801-1795
(907) 465-5324

*Arizona*
Department of Environmental Quality
Technical Review of Water and Wastewater Facilities
3033 N. Central Ave.
Phoenix, AZ 85012
(602) 207-4440
(602) 207-2300

*Arkansas*
Department of Health
Division of Environmental Health Protection
4815 W. Markham St., Slot 46
Little Rock, AR 72205
(501) 661-2171

*California*
California does not have state regulations. A list of county Public Health Department phone numbers and addresses can be provided upon request.

*Colorado*
Department of Public Health and Environment
Water Quality Control Division
4300 Cherry Creek Dr. South
Denver, CO 80222-1530
(303) 692-3500

*Connecticut*
Department of Public Health
410 Capitol Avenue
MS #51 SEW
PO Box 340308
Hartford, CT 06134-0308
(860) 509-7296

Environmental Health Services Division
On-Site Sewage Disposal Section
150 Washington St
Hartford, CT 06106
(203) 240-9277

(Sheet 1 of 7)

Table B-1 (Continued)

*Delaware*
Underground Discharges Branch
Division of Water Resources
Department of Natural Resources and Env. Control
89 Kings Highway
PO Box 1401
Dover, DE  19903
(302) 739-4762

*Florida*
Department of Health and Rehabilitative Service
On-Site Sewage Program
1317 Winewood Blvd.
Tallahassee, FL  32399-0700
(904) 488-4070

*Georgia*
Georgia does not have state regulations.  A list of county Public Health Department phone numbers and addresses can be provided upon request.

*Hawaii*
Department of Health
Wastewater Branch
Environmental Management Division
919 Ala Moana Blvd.
Suite 309
Honolulu, HI  96814
(808) 586-4294

*Idaho*
Department of Health and Welfare
Division of Environmental Quality
1410 N. Hilton, 2nd Flr.
Boise, ID  83706
(208) 334-5860

*Illinois*
Department of Public Health
Division of Environmental Health
525 W. Jefferson St.
Springfield, IL  62761
(217) 782-5830

*Indiana*
Department of Health
Division of Sanitary Engineering
1330 W. Michigan St.
PO Box 1964
Indianapolis, IN  46206-1964
(317) 383-6100
(317) 383-6160

*Iowa*
Department of Natural Resources
Wallace State Office Bldg.
900 E. Grand Ave.
Des Moines, IA  50319-0034
(515) 281-7814
(515) 281-6599

(Sheet 2 of 7)

Table B-1 (Continued)

*Kansas*
Department of Health and Environment
Municipal Section Program
Division of the Environment, Bureau of Water
Non-Point Source Section
Forbes Field, Bldg. 283
Topeka, KS 66620-0001
(913) 296-1683

*Kentucky*
Cabinet for Human Resources
Environmental Management Branch
275 E. Main St.
Frankfurt, KY 4062-0001
(502) 564-4856

*Louisiana*
Department of Health and Hospitals
Office of Public Health
Sanitarian Services
PO Box 60630
New Orleans, LA 70160
(504) 568-5181
(504) 568-8343

*Maine*
Department of Human Services
Health Engineering
State House Station 10
Augusta, ME 04333
(207) 287-5684

*Maryland*
Department of Environment
Water Management Administration
2500 Broening Highway
Baltimore, MD 21224
(410) 631-3779

*Massachusetts*
Department of Environmental Protection
Division of Water Pollution Control
One Winter St., 8th Floor
Boston, MA 01018
(617) 292-5837

*Michigan*
Michigan does not have state-level regulations. A list of county Public Health Departments phone numbers can be provided upon request.

*Minnesota*
Pollution Control Agency
Water Quality Division
Nonpoint Source Compliance Section
520 Lafayette Rd.
St. Paul, MN 55155-4194
(612) 296-8762

*Mississippi*
State Department of Health
PO Box 1700
Jackson, MS 39215-1700
(601) 960-7690

(Sheet 3 of 7)

Table B-1 (Continued)

*Missouri*
Department of Health
PO Box 570
Jefferson City, MO  65102-0570
(314) 751-5937

*Montana*
Department of Environmental Quality
Permit and Compliance Division
Metcalf Building
PO Box 200901
Helena, MT  59620-0901
(406) 444-2825
(406) 444-5344

*Nebraska*
Department of Environmental Quality
Ground Water Section
PO Box 98922
Lincoln, NE  68509-8922
(402) 471-2580

*Nevada*
State Health Division
Bureau of Health Protection Services
505 E. King St., Rm. 103
Capital Complex
Carson City, NV  89710
(702) 687-6353

*New Hampshire*
Department of Environmental Services
Bureau of Wastewater Treatment
6 Hazen Dr.
PO Box 95
Concord, NH  03301
(603) 271-3711

*New Jersey*
Department of Environmental Protection
Division of Water Quality
Bureau of Operational Ground Water Permits
Box CN029
Trenton, NJ  08625-0029
(609) 292-0407

*New Mexico*
    Environment Department
    Drinking Water and Community Services Bureau
    Liquid Waste Program
    525 Camino De Los Marquez Suite #4
    Santa Fe, NM  87502
    (505) 827-7536

    Liquid Waste Program
    Ground Water Bureau
    PO Box 26110
    Santa Fe, New Mexico  87502
    (505) 827-2788

(Sheet 4 of 7)

Table B-1 (Continued)

*New York*
Department of Health
Bureau of Community Sanitation and Food Protection
Room 404
2 University Place
Albany, NY 12203-3399
(518) 458-6706

*North Carolina*
Dept. of Environment, Health and Natural Resources
Division of Environmental Health
ON-Site Wastewater Section
PO Box 27687
Raleigh, NC 27611-7687
(919) 733-2895

*North Dakota*
State Plumbing Board
204 W. Thayer Ave.
Bismark, ND 58501
(701) 328-9977
(701) 221-5210

*Ohio*
Department of Health
Receiving Department (PWS and HSD)
246 N. High St.
Columbus, OH 43266-0588
(614) 466-1390

*Oklahoma*
Department of Environmental Quality
Water Quality Division (ECLS)
1000 NE 10th St.
Oklahoma City, OK 73117-1212
(405) 271-7363
(405) 271-5205

*Oregon*
Department of Environmental Quality
Water Quality Division
811 SW 6th Ave.
Portland, OR 97204-1390
(503) 229-6443

*Pennsylvania*
Department of Environmental Resources
Division of Municipal Planning and Finance
Rachael Carson State Office Building
400 Market St., 10th Floor
Harrisburg, PA 17101-2301
(717) 787-3481
(717) 783-1530

*Rhode Island*
Department of Environmental Management
Division of Groundwater and ISDS
ISDS Section
291 Promenade St.
Providence, RI 02908-5767
(401) 277-4700 Ext. 7700

(Sheet 5 of 7)

Table B-1 (Continued)

*South Carolina*
Department of Health and Environmental Control
Division of Onsite Wastewater Management
Bureau of Environmental Health
2600 Bull St.
Columbia, SC 29201
(803) 935-7835

*South Dakota*
Dept. of Environment and Natural Resources
Air and Surface Water Program
Joe Foss Bldg.
523 E. Capitol
Pierre, SD 57501
(605) 773-3351

*Tennessee*
Dept. of Environment and Conservation
Division of Groundwater Protection
Tenth Floor L & C Tower
401 Church St.
Nashville, TN 37243-1540
(615) 532-0761

*Texas*
Natural Resource Conservation Commission
PO Box 13087
Austin, TX 78711-3087
(512) 239-1000

*Utah*
Department of Environmental Quality
Division of Water Quality
288 N. 1460 West
PO Box 144870
Salt Lake City, UT 84114-4870
(801) 538-6146

*Vermont*
Wastewater Management Division
103 S. Main St.
Sewing Bldg.
Waterbury, VT 05671-0405
(802) 241-3027
(802) 241-3822

*Virginia*
Office of Environmental Health Services
Main Street Station
Suite 117
PO Box 2448
Richmond, VA 23218-2448
(804) 786-1750
(804) 786-3559

*Washington*
State Department of Health
Community Environmental Health Programs LD-11
Building 2, Airdustrial Center
PO Box 47826
Olympia, WA 98504-7826
(360) 586-8133
(360) 586-1249

(Sheet 6 of 7)

**Table B-1 (Concluded)**

*West Virginia*
Office of Environmental Health
Public Health Sanitation Division
815 Quarrier St.
Suite 418
Charleston, WV  25301-2616
(304) 558-2981
(304) 725-0348

*Wisconsin*
Department of Industry, Labor and Human Relations
201 E. Washington Ave. Room 141
Madison, WI  53707
(608) 366-0056

*Wyoming*
Department of Environmental Quality
Water Quality Division
Herschler Building
122 West 25th St.
Cheyenne, WY  82002
(307) 777-7075

(Sheet 7 of 7)

**Table B-2**
**State Regulatory Requirements (NSFC-4, NSFC-5, NSFC-6, NSFC-5, and NSFC-8)**

| State | NPDES State | Application Rate & Fields Sizing | Graywater Systems | No-Flow Toilet | Percolation Tests | Septic Tanks |
|---|---|---|---|---|---|---|
| | | Specific Area | | | | |
| Alabama | Yes | ✓ | ✓ | ✓ | ✓ | ✓ |
| Alaska | No | ✓ | | | | ✓ |
| Arizona | No | ✓ | ✓ | | ✓ | ✓ |
| California | Yes | | | | | |
| Colorado | No | ✓ | | ✓ | ✓ | ✓ |
| Connecticut | Yes | ✓ | ✓ | ✓ | ✓ | ✓ |
| Delaware | No | ✓ | | | ✓ | ✓ |
| Florida | No | ✓ | ✓ | ✓ | | ✓ |
| Georgia | Yes | | | | | |
| Hawaii | Yes | ✓ | ✓ | ✓ | | ✓ |
| Idaho | No | ✓ | | | | ✓ |
| Illinois | Yes | ✓ | | ✓ | ✓ | ✓ |
| Indiana | Yes | ✓ | | | | ✓ |
| Iowa | Yes | ✓ | | ✓ | ✓ | ✓ |
| Kansas | Yes | ✓ | | | ✓ | ✓ |
| Kentucky | Yes | ✓ | ✓ | | ✓ | ✓ |
| Louisiana | No | ✓ | | ✓ | ✓ | ✓ |
| Maine | No | ✓ | ✓ | ✓ | | ✓ |
| Maryland | Yes | ✓ | | ✓ | ✓ | ✓ |
| Massachusett | No | ✓ | | ✓ | ✓ | ✓ |
| Michigan | Yes | | | | | |
| Minnesota | Yes | ✓ | ✓ | ✓ | ✓ | ✓ |
| Mississippi | Yes | ✓ | | ✓ | | ✓ |
| Missouri | Yes | ✓ | | | ✓ | ✓ |
| Montana | Yes | ✓ | | | ✓ | ✓ |
| Nebraska | Yes | ✓ | | | ✓ | ✓ |
| New | No | ✓ | | | ✓ | ✓ |
| New Jersey | Yes | ✓ | ✓ | | ✓ | ✓ |
| New Mexico | No | ✓ | | | ✓ | ✓ |
| New York | Yes | ✓ | ✓ | ✓ | ✓ | ✓ |
| North Carolina | Yes | ✓ | | ✓ | | ✓ |
| North Dakota | Yes | ✓ | | ✓ | ✓ | ✓ |
| Ohio | Yes | | | | | ✓ |
| Oklahoma | No | ✓ | | | ✓ | ✓ |
| Oregon | Yes | ✓ | ✓ | ✓ | | ✓ |
| Pennsylvania | Yes | ✓ | | ✓ | ✓ | ✓ |
| Rhode Island | Yes | ✓ | | ✓ | | ✓ |
| South Carolina | Yes | ✓ | | ✓ | | ✓ |
| South Dakota | No | ✓ | ✓ | ✓ | ✓ | ✓ |
| Tennessee | Yes | ✓ | | | ✓ | ✓ |
| Texas | No | ✓ | ✓ | | ✓ | ✓ |
| Utah | Yes | ✓ | | | ✓ | ✓ |
| Vermont | Yes | ✓ | | | ✓ | ✓ |
| Virginia | Yes | ✓ | | ✓ | ✓ | ✓ |
| Washington | No | ✓ | | ✓ | | ✓ |
| West Virginia | Yes | ✓ | | | ✓ | ✓ |
| Wisconsin | Yes | ✓ | | | ✓ | ✓ |
| Wyoming | Yes | ✓ | ✓ | ✓ | ✓ | ✓ |

**Table B-3**
**States With Specific Design Criteria For Wastewater Treatment System (NSFC-8)**

| State | Treatment Category* | | | | | | | |
|---|---|---|---|---|---|---|---|---|
| | Pre-Treatment | Primary Treatment | Secondary Treatment | Activated Sludge | Ponds and Lagoons | Disinfection | Sludge Treatment and Management | On-Site Systems |
| Alabama | ✓ | | | | | | | ✓ |
| Alaska | | | | | | | | ✓ |
| Arizona | ✓ | ✓ | ✓ | ✓ | | ✓ | ✓ | ✓ |
| Arkansas | | | | | | | | ✓ |
| California | ✓ | ✓ | | | | | | ✓ |
| Colorado | ✓ | ✓ | | ✓ | ✓ | ✓ | | ✓ |
| Connecticut | ✓ | ✓ | ✓ | ✓ | ✓ | ✓ | ✓ | ✓ |
| Delaware | | | | | | | | ✓ |
| Florida | | | | | | | | ✓ |
| Georgia | | | | | | | | ✓ |
| Hawaii | | | | | | | | ✓ |
| Idaho | | | | | | | | ✓ |
| Illinois | ✓ | ✓ | ✓ | ✓ | ✓ | ✓ | ✓ | ✓ |
| Indiana | | | | | | | | ✓ |
| Iowa | ✓ | ✓ | ✓ | ✓ | ✓ | ✓ | ✓ | ✓ |
| Kansas | | ✓ | ✓ | ✓ | ✓ | ✓ | ✓ | ✓ |
| Kentucky | | | | | | | | ✓ |
| Louisiana | | | | | | | | ✓ |
| Maine | ✓ | ✓ | ✓ | ✓ | ✓ | ✓ | ✓ | ✓ |
| Maryland | ✓ | ✓ | ✓ | ✓ | ✓ | ✓ | ✓ | ✓ |
| Massachusetts | ✓ | ✓ | ✓ | ✓ | ✓ | ✓ | ✓ | ✓ |
| Michigan | | | | | | | | ✓ |
| Minnesota | | ✓ | ✓ | ✓ | ✓ | | | ✓ |
| Mississippi | | | | | | | | ✓ |
| Missouri | | | | | | | | ✓ |
| Montana | | | | | | | | ✓ |
| Nebraska | | | | | | | | ✓ |
| Nevada | | | | | | | | ✓ |
| New Hampshire | ✓ | ✓ | ✓ | ✓ | ✓ | ✓ | ✓ | ✓ |
| New Jersey | ✓ | ✓ | ✓ | ✓ | ✓ | ✓ | ✓ | ✓ |
| New Mexico | ✓ | ✓ | ✓ | ✓ | ✓ | ✓ | ✓ | ✓ |
| New York | | | | | | | | ✓ |
| North Carolina | | | | | | | | ✓ |
| North Dakota | | | | | | | | ✓ |
| Ohio | | | | | | | | ✓ |

* See Table B-4 for definitions of treatment categories.

(Continued)

Table B-3 (Concluded)

| State | Pre-Treatment | Primary Treatment | Secondary Treatment | Activated Sludge | Ponds and Lagoons | Disinfection | Sludge Treatment and Management | On-Site Systems |
|---|---|---|---|---|---|---|---|---|
| Oklahoma | ✓ | ✓ | ✓ | ✓ | ✓ | ✓ | ✓ | ✓ |
| Oregon | | | | | | | | ✓ |
| Pennsylvania | ✓ | ✓ | ✓ | ✓ | ✓ | ✓ | ✓ | ✓ |
| Rhode Island | | | | | | | | ✓ |
| South Carolina | | | | | | | | ✓ |
| South Dakota | | | | | ✓ | ✓ | | ✓ |
| Tennessee | ✓ | ✓ | ✓ | ✓ | ✓ | ✓ | ✓ | ✓ |
| Texas | ✓ | ✓ | ✓ | ✓ | ✓ | ✓ | ✓ | ✓ |
| Utah | ✓ | ✓ | ✓ | ✓ | ✓ | ✓ | ✓ | ✓ |
| Vermont | ✓ | ✓ | ✓ | ✓ | ✓ | ✓ | ✓ | ✓ |
| Virginia | ✓ | ✓ | ✓ | ✓ | ✓ | ✓ | ✓ | ✓ |
| Washington | ✓ | ✓ | ✓ | ✓ | ✓ | ✓ | | ✓ |
| West Virginia | ✓ | ✓ | ✓ | ✓ | ✓ | ✓ | ✓ | ✓ |
| Wisconsin | ✓ | ✓ | ✓ | ✓ | ✓ | ✓ | ✓ | ✓ |
| Wyoming | ✓ | ✓ | ✓ | ✓ | ✓ | ✓ | ✓ | ✓ |

**Table B-4**
**Categories in Wastewater Treatment Plants as Defined by State Design Criteria Shown in Table B-3**

| Treatment Category | State Definition |
|---|---|
| Pre-Treatment | Major categories of pre-treatment processes include bar screens, grit removal facilities, and flocculation. |
| Primary Treatment | Major categories of primary treatment included primary clarifiers and chemical coagulation (in settling tanks). |
| Secondary Treatment | The three categories of secondary treatment include attached growth, activated sludge, and ponds and lagoons. |
| Activated Sludge | The two main regulated components of activated sludge are aeration and secondary settling tanks. Design parameters for activated sludge systems are often dependent upon the specific type of activated sludge system utilized; key parameters typically regulated include aeration time, space loading, MLSS concentration, loading factor, sludge age, recirculation factor, side water depth, and surface overflow rate. |
| Ponds and Lagoons | Major categories of ponds and lagoons include stabilization ponds (subcategorized into primary cells and total pond), aerated ponds systems, and aerated facultative lagoons. Key parameters typically regulated include surface loading rate, design loading, water depth, detention time, and the physical dimension of the pond and dike. |
| Disinfection | The sole disinfection category considered is chlorination. Key parameters typically regulated include contact time and dosing capacity for a multitude of conditions (e.g., raw sewage, primary sedimentation effluent, trickling filter plant effluent, etc.). For those states that regulate the same parameters, there is close consensus on values. |
| Sludge Treatment and Management | Major categories of sludge treatment and management include Imhoff tanks, anaerobic digestion, aerobic digestion, heat conditioning, composting, sludge thickening (gravity thickening and dissolved air flotation), and sludge dewatering (vacuum filtration, sludge drying and incineration, sludge drying beds, and shallow sludge lagoons). Key parameters typically regulated include solids loading rate, side water depth, and physical dimensions of a specific technology. |
| On-Site Systems | This section contains the state summary table for on-site systems. Within this general category are the subcategories of septic tanks, distribution boxes, aerobic biological systems, subsurface trench or bed, low pressure distribution systems, siphons, absorption or seepage pits, sand filters, elevated sand mounds, wastewater ponds, and vault privies. |

# Appendix C
# Wastewater Characterization Data

**Table C-1**
**Visitation Data From USACE Recreational Area (Francingues 1976, Middleton USACE)**

| Parameter | Min. Value | Max. Value | Range |
|---|---|---|---|
| Number of people per site | 2 | 6 | 4 |
| Length of stay (day) | 1 | 3 | 2 |

| Size of Camping Party (People) | Frequency of Occurrence | Length of Stay (Days) | Frequency of Occurrence |
|---|---|---|---|
| 1 | 0 | 1 | 19 |
| 2 | 17 | 2 | 22 |
| 3 | 15 | 3 | 36 |
| 4 | 28 | 4 | 15 |
| 5 | 22 | 5 | 7 |
| 6 | 14 | 6 | 3 |
| 7 | 8 | 7 | 3 |
| 8 | 3 | 8 | 2 |
| 9 | 2 | 9 | 1 |
| 10 | 2 | 10 | 3 |
| 11 | 0 | 11 | 0 |
| 12 | 0 | 12 | 0 |
| 13 | 1 | 13 | 0 |
| 14 | 1 | 14 (limit) | 3 |
| 15 | 1 | | |

Note: The average size of camping party was calculated to be 4.78 campers per party.
The average length of stay was calculated to be 3.53 days per party.

**Table C-2**
**Factors Contributing to Water Usage**

| Overnight | Day Use |
|---|---|
| Camping (trailer, houseboat, tent)<br>    Water hookups<br>    Hydrants for tank filling<br>    Pad for trailers, boat docks, and pier for boats<br>Primitive camping (water limited)<br>Other facilities<br>    Hotels, motels<br>    Cabins<br>    Trailers parked in quasi-permanent manner | Picnic areas<br>Comfort stations<br>    Privy vaults<br>    Chemical toilets<br>    Low- and high-volume flush toilets<br>Washhouse, shower, change room<br>Swimming pools<br>Fish-cleaning stations<br>    Spigots<br>    Scoffol disposal<br>Water hydrants<br>Drinking fountains<br>Sprinkler for irrigation |
| **Camping Facilities That Lead To Water Use** | **Miscellaneous Buildings** |
| Comfort station<br>    Privy vault<br>    Chemical toilets<br>    Low- and high-volume flush toilets<br>Washroom shower change room<br>Laundry facilities (sinks, washers)<br>Sanitary disposal stations<br>    Trailer<br>    Marinas<br>Fish-cleaning stations<br>    Spigots<br>    Scoffol disposal<br>Swimming pool<br>Water hydrant | Headquarters<br>Ranger's office<br>Visitor center<br>Officials' homes<br>Janitorial building |

Table C-3
Design Criteria Information (Metcalf & Eddy 1972)

| | | Range of Values Used | | | | | | | |
|---|---|---|---|---|---|---|---|---|---|
| | | Turnover Rate | | No. Persons per Unit | | Water Quantity L/d/c (gpcd) | | Sewage Quantity L/d/c (gpcd) | |
| Type of Area | Type of Facility | Range | Median | Range | Median | Range | Median | Range | Median |
| Day Use | Picnic (comfort station) | 1-1.5 | 1.5 | 3-6 | 4 | 11.4-37 (3-10) | 18.9 (5) | 7.6-30 (2-8) | 15.2 (4) |
| | Overlook (comfort station) | 1-3 | -- | 4-8 | 4 | 7.6-18.9 (2-5) | 15.2 (4) | 7.6-15.2 (2-4) | 11.4 (3) |
| | Boat launching ramp (comfort station) | -- | 1 | 2-4 | 3 | 7.6-18.9 (2-5) | 15.2 (4) | 7.6-15.2 (2-4) | 11.4 (3) |
| Overnight | Camping (tent) water hydrant | -- | -- | -- | 5 | -- | 95 (25) | -- | 72 (19) |
| | Camping (trailer) water hydrant | -- | 1 | 4-5 | 5 | 52-189 (15-50) | 114 (30) | -- | 95 (25) |
| | Camping comfort station | -- | 1 | 4-5 | -- | -- | 95 (25) | -- | 76 (20) |
| | Camping water hydrant plus washers | -- | -- | 5 | -- | -- | 114 (30) | -- | 95 (25) |
| | Camping water hydrant plus trailer hookups | -- | 1 | -- | -- | -- | 132 (35) | -- | 114 (30) |
| Maintenance | Shops | -- | -- | -- | -- | -- | 18.9 (5) | -- | 15.2 (4) |
| Visitor | Information Center | -- | -- | -- | -- | -- | 7.6 (2) | -- | 3.8 (1) |

Table C-4
Average Wastewater Flows From Recreation Areas (EPA-625-R-92/005, Matherly 1975, and Metcalf & Eddy 1972)

| Source | | Unit | Flow (L/d/unit (gpd/unit)) Typical | Range |
|---|---|---|---|---|
| Apartment, resort | | Person | 220 (58.1) | 200-280 (52.8-74) |
| Bathhouse | | Person | 38 (10) | -- |
| Cabin, resort | | Person | 160 (42.3) | 130-190 (34.3-50.2) |
| Cafeteria | | Customer | 6.0 (1.6) | 4.2-9.5 (1.1-2.5) |
| | | Employee | 41 (10.8) | 30-50 (7.9-13.2) |
| Camps | Camping, tent | Person | 76 (20) | -- |
| | Camping, trailer | Person | 95 (25) | -- |
| | Day camp (no meals) | Person | 49 (13) | -- |
| | Resort, limited plumbing | Person | 189 (50) | -- |
| | Tourist, central bath, and toilet facilities | Person | 132 (35) | -- |
| Cottages (seasonal occupancy) | | Person | 189 (50) | -- |
| Cocktail Lounge | | Seat | 75 (19.8) | 50-100 (13.2-26.4) |
| Coffee Shop | | Customer | 20 (5.3) | 15-30 (4.0-7.9) |
| | | Employee | 41 (10.8) | 30-50 (7.9-13.2) |
| Country Club | | Member present | 401 (106) | 250-500 (66-132) |
| | | Employee | 50 (13.2) | 40-60 (10.6-15.9) |
| Dining Hall | | Meal served | 30 (7.9) | 15-50 (4.0-13.2) |
| Dormitory, bunkhouse | | Person | 150 (39.6) | 75-175 (19.8-46.2) |
| Hotel, resort | | Person | 200 (52.8) | 150-240 (39.6-63.4) |
| Laundromat | | Machine | 2199 (581) | 1802-2525 (476-667) |
| Store, resort | | Customer | 10 (2.6) | 4.9-20 (1.3-5.3) |
| | | Employee | 40 (10.6) | 30-50 (7.9-13.2) |
| Swimming Pool | | Customer | 40 (10.6) | 20-50 (5.3-13.2) |
| | | Employee | 40 (10.6) | 30-50 (7.9-13.2) |
| Theater | | Seat | 9.5 (2.5) | 7.6-15.2 (2.0-4.0) |
| Visitor Center | | Visitor | 20 (5.3) | 15.2-30 (4.0-7.9) |
| Parks | Overnight, flush toilet | Person | 76 (20) | -- |
| | Trailers, individual bath | Person | 189 (50) | -- |
| Picnics | Bathhouse, showers, and flush toilets | Person | 76 (20) | -- |
| | Toilet, only | Person | 15 (4) | -- |
| | Vault | Person | 1.1 (0.28) | -- |
| Marinas | Toilet, only | Person | 38 (10) | -- |
| | Toilet and Shower | Person | 60 (16) | -- |

**Table C-5**
**Typical Rates of Water Use For Various Devices (Corbitt 1990)**

| Device | | Range of Flow | |
|---|---|---|---|
| | | SI Units | Customary Units |
| Automatic home laundry machine | | 110-200 L/load | 39-53 gal/load |
| Automatic home-type dishwasher | | 15-30 L/load | 4-8 gal/load |
| Automatic home-type washing machine | | 130-200 L/use | 34-53 gal/use |
| Bathtub | | 90-110 L/use | 24-29 gal/use |
| Continuous-flow drinking fountain | | 4-5 L/min | 1 gal/min |
| Dishwashing machine, commercial*: | Conveyor type, at 100 kN/m$^2$ | 15-25 L/min | 4-7 gal/min |
| | Stationary rack type, at 100 kN/m$^2$ | 25-35 L/min | 7-9 gal/min |
| Fire hose, 36 mm, 13-mm nozzle, 20-m head | | 140-160 L/min | 37-42 gal/min |
| Garbage disposal unit, home-type | | 6000-7500 L/wk | 1600-2000 gal/wk |
| Garden grinder | | 4-8 L/person/day | 1-2 gal/person/day |
| Garden-hose, 16 mm, 8-m head | | 10-12 L/min | 3 gal/min |
| Garden hose, 19 mm, 8-mm head | | 16-20 L/min | 4-5 gal/min |
| Lawn sprinkler | | 6-8 L/min | 2 gal/use |
| Lawn sprinkler, 280 m$^2$ lawn, 25 mm/wk | | 6000-7500 L/wk | 1600-2000 gal/wk |
| Shower head, 16 mm, 8-m head | | 90-110 L/use | 24-29 gal/use |
| Washbasin | | 4-8 L/use | 1-2 gal/use |
| Water closet, flush valve, 170 kN/m$^2$ | | 90-110 L/min | 24-29 gal/use |
| Water closet, tank | | 15-25 L/use | 4-7 gal/use |

\* Does not include water to fill wash tank.

Table C-6
Minimum Required Number of Sanitary Fixtures (Penn Bureau of Resources and USDOI 1958)

| | Types of Fixtures | | | | | | | |
| --- | --- | --- | --- | --- | --- | --- | --- | --- |
| | Water Closets | | | Urinals[4] | | Lavatories | | |
| Types of Building Occupancy | Number of Sites | No. of Fixtures[3] Male | Female | Number of Sites | No. of Fixtures[3] | Number of Sites | No. of Fixtures[3] Male | Female |
| Comfort stations[1] for campgrounds | 1-20 | 1 | 2 | 1-20 | 1 | 1-20 | 1 | 1 |
| | 21-30 | 2 | 3 | 21-30 | 2 | 21-30 | 2 | 2 |
| | Number of Car Parking Spaces | No. of Fixtures[3] Male | Female | Number of Car Parking Spaces | No. of Fixtures[3] | Number of Car Parking Spaces | No. of Fixtures[3] Male | Female |
| Comfort stations[1] for picnic areas | 1-40 | 1 | 2 | 1-40 | 1 | 1-40 | 1 | 1 |
| | 41-80 | 2 | 4 | 41-80 | 2 | 41-80 | 2 | 2 |
| | 81-120 | 3 | 6 | 81-120 | 3 | 81-120 | 3 | 3 |

| | Number of Seasonal Slips | No. of Commodes Male | Female | No. of Urinals | No. of Lavatories Male | Female | No. of Showers Male | Female |
| --- | --- | --- | --- | --- | --- | --- | --- | --- |
| Comfort stations for marinas[2] | 0-49 | 1 | 1 | 0 | 1 | 1 | 0 | 0 |
| | 50-99 | 1 | 2 | 1 | 1 | 1 | 0 | 0 |
| | 100-149 | 2 | 3 | 1 | 2 | 2 | 1 | 1 |
| | 150-199 | 2 | 4 | 2 | 3 | 3 | 2 | 2 |
| | 200-249 | 3 | 5 | 2 | 4 | 4 | 2 | 2 |
| | Number of Transient Slips | | | | | | | |
| | 0-24 | 1 | 1 | 1 | 1 | 1 | 1 | 1 |
| | 25-49 | 1 | 2 | 1 | 2 | 2 | 2 | 2 |
| | 50-74 | 2 | 3 | 1 | 2 | 2 | 2 | 2 |
| | 75-100 | 2 | 4 | 2 | 3 | 3 | 3 | 3 |

1  A comfort station should contain 2 water closets, 2 lavatories, and 1 urinal for males, and 3 water closets and 2 lavatories for females. A sani-stand substituted for 1 water closet is desired for females. A comfort station should provide facilities for 60 percent of users inside a 91-meter (300-foot) radius, 90 percent inside a 137-meter (450-foot) radius, and 100 percent inside a 183-meter (600-foot) radius except where conformance with local health codes might require a smaller radius.

2  For marinas comfort stations where piped water is available, the facility shall consist of a minimum of one commode and one lavatory for each male and female for every 100 seasonal slips. Sanitary facilities may consist of privys where piped water is not available.

3  All fixtures are subject to possible malfunction and vandalism; therefore, stand-by fixtures should be provided even though user ratio would require only one fixture.

4  Drinking fountains should not be installed in toilet rooms. Where urinals are provided for women, the same number should be provided as for men.

Table C-7
**Calculation of Design Flow For Fixture Unit Method (Penn Bureau of Resources and USDOI 1958)**

| Item | Rate L/hr/fixture (gal/hr/fixture) | Duration (hr/day) | | |
|---|---|---|---|---|
| | | Camping | Picnic | Fishing |
| Water Closet | 136 (36) | 8 | 10 | -- |
| Water Fountain | 38 (10) | 4 | 8 | -- |
| Urinal | 38 (10) | 8 | 10 | -- |
| Laundry | 189 (50) | 4 | -- | -- |
| Lavatory | 57 (15) | 8 | 10 | -- |
| Dump Station | 38 (10) | 4 | -- | -- |
| Shower | 380 (100) | 3 | -- | -- |
| Fish Cleaning Station | 189 (50) | -- | -- | 4 |
| Service Sink | 38 (10) | 2 | 2 | -- |

Table C-8
**Monthly Flow Distribution**

| Month | Percent of Peak Flow |
|---|---|
| January-March | 20 |
| April | 21 |
| May | 42 |
| June-August | 100 |
| September | 42 |
| October | 21 |
| November-December | 20 |

Table C-9
Recreation Areas and Domestic Wastewater Characteristics (Francingues 1976, Matherly 1975, and Metcalf & Eddy 1972)

| Parameter | Waste Source Number [a] | | | | | |
|---|---|---|---|---|---|---|
| | #1 | #2 | #3 | #4 | #5 | #6 |
| TOC (mg/L) | 120 | 158 | 144 | 1980 | -- | 200 |
| BOD (mg/L) | 203 | 229 | 196 | 3320 | 145 | 200 |
| COD (mg/L) | 430 | 439 | 336 | 6370 | 388 | 500 |
| BOD/TOC | 1.69 | 1.45 | 1.36 | 1.68 | -- | 1 00[b] |
| COD/TOC | 3.58 | 2.78 | 2.33 | 3.22 | -- | 2 50[b] |
| Total P (mg/L) | -- | 10.6 | 10.3 | 166 | 10 | 10 |
| Ortho-PO$_4$ (mg/L) | -- | 8.0 | 7.5 | 72 | 8 | 7 |
| TKN (mg/L) | 48.1 | 44.9 | 82.8 | 1040 | 114 | 40 |
| NH$_3$ (mg/L) | 39.4 | 40.5 | 82.4 | 1000 | 64 | 25 |
| Cl[c] | -- | 84.4 | 91.9 | 1070 | -- | 50 |
| pH range (S. U.) | -- | (7.2-8.7) | (7.3-8.3) | (8.0-8.3) | -- | -- |
| Alkalinity (mg/L) | 356 | -- | -- | -- | -- | 100 |
| TS (mg/L) | 613 | 680 | 634 | 9100 | -- | 700 |
| TVS (mg/L) | 332 | 352 | -- | -- | -- | 250 |
| TSS (mg/L) | 252 | 281 | 285 | 2860 | -- | 200 |
| TVSS (mg/L) | 161 | 185 | -- | -- | -- | 50 |
| Settleable solids (mg/L) | 17 | -- | -- | -- | -- | 10 |

Note: (a)  #1 denotes camping area without trailer dump waste.
      #2 denotes camping area with trailer dump waste.
      #3 denotes picnic area waste.
      #4 denotes picnic area waste.
      #5 denotes average of three combination picnic and camping areas, Shelbyville, IL
      #6 denotes typical domestic waste, medium concentraiton (Metcalf & Eddy 1972 and USAEWES)
   (b)  Slightly lower than values in other principal tests.
   (c)  Should be adjusted for Cl in supply water.

Table C-10
Characteristics of a 3 785 L (1 000 gal) load of Nonwater Carriage Wastes (Smith 1973)

| | BOD kg (lb) | Solids kg (lb) | Volatile Solids kg (lb) | Chemical Formulations | |
| | | | | Formaline Forest Area kg (lb) | Zinc Marine kg (lb) |
|---|---|---|---|---|---|
| Septage | 26 (58) | 136 (300) | 102 (225) | -- | -- |
| Vault wastes[a] | 73 (160) | 152 (335) | 146 (320) | -- | -- |
| Low-volume wastes | 57 (125) | 118 (260) | 68 (150) | -- | -- |
| Recirculating and portable chemical toilet wastes[a] | 73 (160) | 152 (335) | 146 (320) | 2.3-4.6 (5-10) | 2.3-4.6 (5-10) |

(a) Assuming waste strength diluted by half to facilitate pumping and cleaning

Table C-11
Characteristics of Vault Wastes at CE Facility (Harrison 1972 and Simmons 1972)

| Parameter | Vault Wastes With Short Detention Times and no Chemical Additives | | | Vault Wastes With Longer Detention Times and Chemical Additives | | | Standard Vault (Nonleaking)[a, b] |
| | Min. | Avg. | Max. | Min. | Avg. | Max. | |
|---|---|---|---|---|---|---|---|
| BOD, mg/L | 5900 | 8780 | 12 900 | 390 | 3895 | 9450 | 20 000 |
| Total solids, mg/L | 6520 | 9675 | 20 080 | 2022 | 10 169 | 21 292 | 55 000 |
| Suspended solids, mg/L | 700 | 3660 | 10 680 | 610 | 2667 | 6450 | 25 000 |
| Dissolved solids, mg/L | 3628 | 6015 | 10 640 | 570 | 7001 | 12 080 | --- |
| pH, S.U. | --- | --- | --- | 7.1 | --- | 8.5 | --- |
| COD, mg/L | --- | --- | --- | --- | --- | --- | 40 000 |

Note: (a) Standard vaults with chemcial or oil-recirculation toilet--use same values.
(b) Standard vaults with low-volume flush toilet--use 25% of values

**Table C-12**
**Estimated Undiluted Vault Waste Characteristics (USAEWES)**

| Parameter | Estimated Values |
|---|---|
| $BOD_5$, mg/L | 2640 |
| COD, mg/L | 9100 |
| Total solids, mg/L | 11 180 |
| Suspended solids, mg/L | 4150 |
| Dissolved solids, mg/L | 8060 |
| Total volatile solids, mg/L | 4340 |
| Total volatile suspended solids, mg/L | 220 |
| Total P, mg/L | 230 |
| $O-PO_4$, mg/L | 290 |
| Total N, mg/L | 890 |
| $NO_2$-N, mg/L | <0.05 |
| $NO_3$-N, mg/L | 590 |
| Total organic carbon, mg/L | 1120 |
| $CO_3$ (as $CaO_3$), mg/L | 110 |
| $HCO_3$ (as $CaO_3$), mg/L | 2880 |
| pH, S.U. | 8.5 |
| Total coliforms (MPN/100 mL) | $2.4 \times 10^5$ |

Table C-13
Wastewater Characteristics of Dump Stations (USEPA-1)

| Characteristics | #1 | #2 | #3 | #4 | #5 | #6 | #7 | #8 | #9 | #10 | #11 | Range |
|---|---|---|---|---|---|---|---|---|---|---|---|---|
| | | | | | | Sample Number | | | | | | |
| Total solids, wt % | -- | 1.90 | 1.33 | 0.21 | 0.92 | 1.11 | 1.85 | -- | 0.51 | -- | -- | 0.21-4.40 |
| Suspended solids, wt % | -- | 0.98 | 1.03 | 0.04 | 0.65 | 0.55 | 1.20 | -- | 0.20 | -- | -- | 0.04-5.85 |
| Filterable residue, wt % | -- | 0.92 | 0.30 | 0.17 | 0.27 | 0.56 | 0.65 | -- | 0.31 | -- | -- | 0.17-6.60 |
| Fixed residue, wt % | -- | 0.81 | 0.55 | 0.11 | 0.47 | 0.41 | 1.64* | -- | 0.26 | -- | -- | 0.11-5.80 |
| pH, S.U. | 8.0 | 8.6 | 8.2 | 7.0 | 6.0 | 6.1 | 6.6 | 8.3 | 8.3 | 7.6 | 6.4 | 6.0-8.6 |
| Conductivity, $\mu$ ohms/cm | 40 000 | 26 000 | 12 000 | 1400 | 4410 | 10 800 | 9900 | 9800 | 13 000 | 42 000 | -- | 1400-42 000 |
| COD, mg/L | 2576 | 20 360 | 16 500 | 9900 | 24 400 | 24 400 | 42 800 | 38 700 | 4680 | 1296 | 102 650* | 1296-102 650 |
| Soluble COD, mg/L | 1240 | 7008 | 3760 | 9900 | 1820 | 5440 | 9680 | 2300 | 3600 | 634 | 23 600 | 684-23 600 |
| Total chromium, mg/L | <0.06 | <0.06 | <0.06 | <0.06 | <0.06 | <0.06 | <0.06 | <0.06 | <0.06 | <0.06 | <0.06 | <0.06 |
| Hexavalent chromium, mg/L | <0.06 | <0.06 | <0.06 | <0.06 | <0.06 | <0.06 | <0.06 | <0.06 | <0.06 | <0.06 | <0.06 | <0.06 |
| Copper, mg/L | <0.1 | <0.1 | <0.1 | <0.1 | <0.1 | <0.1 | <0.1 | <0.1 | <0.1 | <0.1 | <0.1 | <0.1 |
| Zinc, mg/L | <1.0 | <1.0 | <1.0 | 2.6 | 34 | 181* | 79 | 28 | <1.0 | <1.0 | 8.5 | <1.0-181 |
| Nickel, mg/L | <0.1 | <0.1 | <1.0 | <0.1 | <0.1 | <0.1 | <0.1 | <0.1 | <0.1 | <0.1 | <0.01 | <0.01-0.1 |
| Chlorides, mg/L | 13 490* | 4383 | 2350 | <2.0 | 6.0 | <2.0 | <2.0 | <2.0 | 1821 | 11 780 | 30 | <2.0-13 490 |
| Sulfate, mg/L | 860 | 1050 | 220 | 190 | 1650 | 713 | 710 | 650 | 330 | 1350 | 4125* | 190-4125 |
| Total phosphorous, mg/L | 12 | 350 | 3956 | 23 | 138 | 164 | 166 | 138 | 1620 | 85 | 1625 | 12-1625 |
| Total Kjeldahl Nitrogen, mg/L | 4285 | 8487* | 100 | 506 | 467 | 2317 | 1803 | 1695 | 2303 | 1804 | -- | 467-8487 |
| Total organic nitrogen, mg/L | 907* | 100 | 100 | 84.2 | 56.9 | 325 | 222 | 273 | 3.0 | 4.1 | -- | 3.0-907 |
| Total inorganic nitrogen, mg/L | 3378 | 8500 | 22.0 | 421.7 | 410 | 1992 | 1 581 | 1422 | 2300 | 1930 | 4000 | 410-8500 |
| Nitrates-nitrogen, mg/L | 0.5 | 13.0 | <0.1 | 1.7 | 2.0 | 12 | 1.0 | 22 | <1.0 | 130* | <1.0 | 0.5-130 |
| Nitrites-nitrogen, mg/L | <0.1 | <0.1 | 3956 | <0.1 | <0.1 | <0.1 | <0.1 | <0.1 | <0.1 | <0.1 | <0.1 | <0.1 |
| Ammonia nitrogen, mg/L | 3378 | 8847* | 3956 | 420 | 408 | 1980 | 1580 | 1400 | 2300 | 1800 | 4000 | 408-8847 |
| Anionic detergents, mg/L | <0.05 | 2.4 | <0.05 | 30* | <0.05 | <0.05 | 3.2 | 3.0 | 5.6 | <1.0 | 4000 | <0.05-30 |
| Iodine and iodides mg/L | 0.02 | 0.02 | 0.02 | <0.02 | <0.02 | <0.02 | <0.02 | <0.02 | <0.02 | <0.02 | <0.02 | <0.02 |

Note: Sample #4 not analyzed because of excessive amount of oil in sample
* The values are considered outliers based on the Dixon method at the 95% confidence level. Average values should be calculated without the outliers (AOAC 1982).

**Table C-14**
**Waste Characterization from Watercraft Pumpage (Robin and Green)**

| Parameter | Concentration Value |
|---|---|
| TSS, mg/L | 1400-2400 |
| TOC, mg/L | 1500-3900 |
| $BOD_5$, mg/L | 1700-3500 |
| COD, mg/L | 4400-7099 |
| Total Nitrogen, mg/L | 1600-2000 |
| Coliforms, MPN/100 mL | $10^2$-$10^{10}$ |
| Zinc, mg/L | 25-250 |

**Table C-15**
**Characteristics of Fish-Cleaning Station Effluent (Matherly 1975)**

| Parameter | Minimum | Mean | Maximum |
|---|---|---|---|
| pH (S. U.) | 6.9 | -- | 8.5 |
| COD (mg/L) | 130 | 200 | 750 |
| $BOD_5$ | 85 | 100 | 500 |
| SS (mg/L) | 15 | 25 | 75 |

**Appendix D**
**Wastewater Design Criteria and Examples Matrix Summary from Non-Military Sources**

Table D-1
Conventional Wastewater Treatment - Preliminary, Sedimentation, and Biological Processes

Reference

| Topic | Theory and Practice of Water and Wastewater Treatment, John Wiley, 1997 | Unit Operations and Processes in Environmental Engineering, PWS Publishing, 1995 | Environmental Science and Engineering, Prentice-Hall, 1996 | Wastewater Engineering: Treatment, Disposal and Reuse, McGraw-Hill, 1991 | Design of Municipal Wastewater Treatment Plants, WEF/ASCE, 1992 | Domestic Wastewater Treatment, TM 5-814-3 |
|---|---|---|---|---|---|---|
| Screens | Chapter 11 (T) | Chapters 6, 7 | -- | Chapter 9 (T) | Chapter 9 (T) | -- |
| Bar Racks | Chapter 11 | Chapter 7 (E) | -- | Chapter 9 (T) | Chapter 9 (T) | -- |
| Barminutors | -- | Chapter 7 | -- | Chapter 9 (T) | Chapter 9 (T) | -- |
| Comminution | Chapter 9 | Chapter 7 | -- | Chapter 9 | Chapter 9 | -- |
| Grit Removal | Chapter 11 (T), (E) | Chapter 7 (T), (E) | -- | Chapter 9 (T) | Chapter 9 (T) | -- |
| Mechanical | Chapter 11 (T), (E) | Chapter 7 (T), (E) | -- | Chapter 9 (T) | Chapter 9 | -- |
| Aerated | Chapter 11 (T) | Chapter 7 (T), (E) | -- | Chapter 9 (T), (E) | Chapter 9 | -- |
| Flow Equalization | Chapter 9, 10 | Chapter 7 (T), (E) | -- | Chapter 9 | Chapter 9 | -- |
| Imhoff Tank | Chapter 18 | -- | -- | Chapter 14 (T) | Chapter 10 (T) | -- |
| Conventional Sedimentation | Chapter 11 (T), (E) | Chapter 9 (T), (E) | Chapter 12 (T), (E) | Chapter 9 (T), (E) | Chapter 10 | -- |
| Chemical Precipitation | -- | -- | -- | -- | Chapter 10 (T) | -- |
| Enhanced Coagulation | -- | -- | -- | -- | Chapter 10 (T) | -- |
| Trickling Filter | Chapter 17 (T) | Chapter 17 (T), (E) | Chapter 12 (T), (E) | Chapters 8, 10 (T), (E) | Chapter 12 (T), (E) | Chapter 7 (E) |
| Oxidation Ditches (CLR) | -- | Chapter 15 (T), (E) | -- | -- | Chapter 11 (T) | Chapter 13 (T), (E) |
| Rotating Biological Contactors | -- | Chapter 17 (E) | -- | Chapter 10 (T), (E) | Chapter 12 (T), (E) | -- |
| Sequenching Batch Reactors | Chapter 17 (E) | Chapter 3 (E) | -- | Chpater 10 (T), (E) | -- | -- |
| Sludge Pumping System | -- | -- | -- | Chapter 12 (T), (E) | Chapter 17 | Chapter 16 (T), (E) |
| Sludge Pumping Scum | -- | -- | -- | Chapter 10 (T) | -- | Chapter 6 |
| Effluent Design | -- | -- | -- | Chapter 10 (T) | -- | Chapter 6 |

(T)    Design criteria table(s) provided
(E)    Design example(s) provided

**Table D-2**
**Sludge Handling, Treatment, and Disposal**

| Topic | Theory and Practice of Water and Wastewater Treatment John Wiley, 1997 | Unit Operations and Processes in Environmental Engineering PWS Publishing, 1995 | Environmental Science and Engineering Prentice-Hall, 1996 | Wastewater Engineering: Treatment, Disposal and Reuse McGraw-Hill, 1991 | Design of Municipal Wastewater Treatment Plants WEF/ASCE, 1992 | Domestic Wastewater Treatment TM 5-814-3 | Dewatering Municipal Wastewater Sludges Design Manual, September 1987 EPA/625/1-87-014 |
|---|---|---|---|---|---|---|---|
| | | | | Reference | | | |
| Anaerobic Sludge Digestion | Chapter 18 (T) | Chapter 19 (T), (E) | Chapter 12 (T), (E) | Chapters 8, 12 (T), (E) | Chapter 18 (T) | Chapter 16 (T), (E) | -- |
| Aerobic Sludge Digestion | Chapter 18 (T) | Chapter 20 (T), (E) | -- | Chapters 8, 12 | Chapter 18 (T) | -- | -- |
| Sludge Handling | -- | -- | -- | -- | -- | -- | -- |
| Sludge Disposal Regulations | -- | -- | -- | -- | -- | -- | -- |
| Sludge Handling, Treatment, and Disposal | Chapter 20 (T), (E) | Chapter 21 (T), (E) | -- | Chapter 12 (T), (E) | Chapter 12 (T) | Chapter 16 (E) | Chapters 3, 5, 6, 7 (T), (E) |
| Sludge: Regulation Considerations | Chapter 10 (T) | Chapters 1, 2, 3, 4, 5 (T), (E) | Chapters 2, 3, 4 (T) | Chapters 4, 5, 6, 9, 10 (T), (E) | Chapter 12 (T) | Chapters 1, 2, 3 (T) | -- |

(T) Design criteria table(s) provided
(E) Design example(s) provided

Table D-3
Small Wastewater Treatment Systems

| Topic | Wastewater Engineering: Treatment, Disposal and Reuse McGraw-Hill, 1991 | Appropriate Technology for Treating Wastewater at Remote Sites, CERL, 1984 | Septic Systems Handbook, 1987 Lewis Publishers | Mound Systems: Design Module Number 9, NSFC, 1990 | Mound Systems Pressure Distribution of Wastewater Ohio State Univ., 1992 | Onsite Wastewater Treatment Systems Hogarth House, 1994 | Design of Municipal Wastewater Treatment Plants WEF/ASCE, 1992 | Onsite Wastewater Treatment and Disposal Systems October 1980 EPA/625/ 1/80/012 |
|---|---|---|---|---|---|---|---|---|
| Septic Tank System | Chapter 14 (T), (E) | Chapter 14 (T), (E) | Chapter 2 | -- | -- | Chapter 7 (T), (E) | -- | Chapter 6 |
| Grease and Oil Interception Tanks | Chapter 14 | -- | -- | -- | -- | -- | -- | -- |
| Imhoff Tanks | Chapter 14 | -- | -- | -- | -- | -- | -- | -- |
| Intermittent Sand Filter | Chapter 14 (T), (E) | Chapter 14 (T), (E) | Chapter 13 | -- | -- | -- | -- | -- |
| Absorption (Disposal) Fields | Chapter 14 (T), (E) | -- | Chapter 2 | -- | -- | Chapter 7 (T) | -- | Chapter 7 (T), (E) |
| Mound System | Chapter 14 | Chapter 14 | -- | Chapter 1 (T), (E) | Chapter 1 (T), (E) | Chapter 7 (T) | -- | Chapter 7 (T), (E) |
| Pit Privy | -- | -- | -- | -- | -- | -- | Chapter 5 | -- |
| Vault Toilets and Aerated Vault Latrines | -- | -- | -- | -- | -- | -- | Chapter 4 | -- |
| Composting Toilets | -- | -- | -- | -- | -- | -- | Chapter 2 | -- |
| Contact Stabilization | Chapter 14 | -- | -- | -- | -- | -- | -- | -- |

(T) Design criteria table(s) provided
(E) Design example(s) provided

(Continued)

**Table D-3 (Concluded)**

Reference

| Topic | Wastewater Engineering: Treatment, Disposal and Reuse McGraw-Hill, 1991 | Wastewater Treatment and Disposal for Small Communities September 1992 EPA/625/R-92/005 | Small Community Water and Wastewater Treatment September 1992 EPA/625/R-92/010 | Mound Systems: Design Module Number 9 NSFC, January 1990 | Mound Systems Pressure Distribution of Wastewater Ohio State Univ., 1992 | Alternative Wastewater Collection Systems October 1991 EPA 625/1-91/024 | Domestic Septage Guidance September 1993 EPA/832/B-92/005 |
|---|---|---|---|---|---|---|---|
| Wastewater Collection Systems | Chapter 11 | -- | -- | -- | -- | -- | -- |
| Pressure Sewers | Chapter 14 (T) | (T) | (T) | -- | -- | Chapters 2, 5 (T), (E) | -- |
| Vacuum Sewers | Chapter 14 (T) | (T) | (T) | -- | -- | Chapters 3, 5 (T), (E) | -- |
| Small Diameter Gravity Sewers | -- | (T) | (T) | -- | -- | Chapters 4, 5 (T), (E) | -- |
| Physical-Chemical Treatment | Chapter 14 | -- | -- | -- | -- | -- | -- |
| Package Plants | Chapter 14 (T), (E) | -- | (T) | -- | -- | -- | -- |
| Lagoons (Ponds) | Chapter 14 | (E) | (E) | -- | -- | -- | -- |
| Land Treatment Methods | Chapter 14 | (E) | (E) | -- | -- | -- | -- |
| Submerged Bed Constructed Wetlands | -- | (E) | (E) | -- | -- | -- | -- |
| Septage Holding Tank or Station | Chapter 14 | -- | -- | -- | -- | -- | -- |
| Septage and Septage Disposal | Chapter 14 (E) | (T) | -- | -- | -- | -- | Chapters 2, 3, 4 (T) |
| Effluent Disposal Options | Chapter 14 | -- | -- | -- | -- | -- | -- |
| Sludge Treatment | -- | (T) | (T) | -- | -- | -- | -- |
| Effluent Treatment | -- | -- | (E) | -- | -- | -- | -- |

T = Design criteria table(s) provided
E = Design example(s) provided

**Table D-4**
**Natural Systems Wastewater Treatment Design Criteria and Examples**

| Topic | Unit Operations and Processes in Environmental Engineering 1995 PWS Publishing | Wastewater Engineering: Treatment, Disposal and Reuse McGraw-Hill, 1991 | Theory and Practice of Water and Wastewater Treatment John Wiley, 1997 | Theory and Practice of Water and Wastewater Treatment, 400/S-90/011 EPA, July 1990 | Natural Systems for Wastewater Treatment WEF 1990 | Constructed Wetlands and Aquatic Plant Systems for Municipal Wastewater Treatment, 625/1-88/022 EPA, Sept. 1988 | Water Treatment and Disposal for Small Communities, 625/R-92/005 EPA, Sept. 1992 | Summary Report Small Community Water and Wastewater Treatment, EPA, Sept. 1992 | Subsurface Flow Constructed Flow Constructed Wetlands for Wastewater Treatment, 832/R-93/008 EPA, July 1993 |
|---|---|---|---|---|---|---|---|---|---|
| Constructed Wetlands - Aquatic Plant Systems | -- | -- | -- | -- | -- | Chapter 4 (T) | -- | -- | -- |
| Constructed Wetlands - Subsurface Flow | -- | -- | -- | -- | -- | -- | -- | -- | Chapter 4 (T) |
| Constructed Wetlands - Wetland Types | -- | -- | -- | Chapter 3 (T) | Chapter 9 (T) | -- | -- | -- | -- |
| Floating Aquatic Plant Treatment | -- | Chapter 13 (T) | -- | -- | Chapter 8 (T) | Chapter 13 (T) | -- | -- | -- |
| Land Application - Overland Flow | Chapter 22 (T), (E) | Chapter 13 (E) | Chapter 19 (T) | -- | -- | -- | Chapter 5 (T) | Chapter 1 (T) | -- |
| Land Application - Rapid Infiltration | -- | -- | -- | -- | -- | -- | Chapter 5 (T) | Chapter 1 (T) | -- |
| Land Application - Slow Rate | Chapter 22 (T), (E) | Chapter 13 (T), (E) | Chapter 19 (T) | -- | -- | -- | Chapter 5 (T) | Chapter 1 (T) | -- |
| Stabilization Ponds or Aerobic Lagoons | Chapter 18 (E) | Chapters 8, 10 (T), (E) | Chapter 19 (T) | -- | -- | -- | -- | -- | -- |
| Stabilization Ponds or Anaerobic Lagoons | -- | Chapters 8, 10 (T), (E) | Chapter 19 (T) | -- | -- | -- | -- | -- | -- |
| Stabilization Ponds or Facultative Lagoons* | -- | Chapters 8, 10 (T), (E) | Chapter 19 (T) | -- | -- | Chapter 13 | -- | -- | -- |
| Subsurface Wastewater Infiltration System | -- | -- | -- | -- | Chapter 3 (T)-- | -- | -- | -- | -- |

(T)  Design criteria table(s) provided
(E)  Design example(s) provided
*  Aerobic - Anaerobic

**Table D-5**
**Effluent Disinfection Design Criteria and Examples**

| Topic | Reference | | | | |
|---|---|---|---|---|---|
| | Theory and Practice of Water and Wastewater Treatment John Wiley, 1997 | Unit Operations in Environmental Engineering PWS Publishing, 1995 | Environmental Science and Engineering Prentice-Hall, 1996 | Wastewater Engineering: Treatment, Disposal and Reuse McGraw-Hill, 1991 | Design of Municipal Wastewater Treatment Plants WEF/ASCE, 1992 |
| Effluent Disinfection-Chlorination | Chapter 16 | Chapter 24 | Chapter 12 | Chapter 7 | Chapter 14 |
| Effluent Disinfection-Ozonation | Chapter 16 | Chapter 24 | Chapter 12 | Chapter 7 | Chapter 14 |
| Effluent Disinfection-Ultraviolet | Chapter 16, (T), (E) | Chapter 24 | Chapter 12 | Chapter 7 | Chapter 14 |

(T)  Design criteria table(s) provided
(E)  Design example(s) provided

## Appendix E
## Design Examples

### E-1. Package Plant Extended Aeration

Design an extended aeration package plant (prefabricated or pre-engineered) to treat a municipal wastewater flow of 125 000 L/d (33 000 gal/d). Solids retention typically ranges from 20 to 30 days; MLSS varies between 3 000 and 6 000 mg/L. The Food to Microorganism ratio (F/M) typically varies between 0.05 to 0.30. Influent $BOD_5$ and TSS will generally be about 250 mg/L. The dissolved oxygen (DO) concentration is in the 1.5 to 2.5 mg/L range and preferably will never be below 2.0 mg/L. Coarse bubble aerators will be used. Detention time in the aeration tank will be one day.

**Table E-1**
**Design Assumptions**

Influent/Effluent Composition - Given

| Parameter | Influent | Effluent |
|---|---|---|
| $BOD_5$ | 250 mg/L | 20 mg/L |
| TSS | 250 mg/L | 20 mg/L |
| TKN | 40 mg/L | 5 mg/L |

*Assumptions*

| | |
|---|---|
| Minimum operating temperature: | 17°C (62°F) |
| Site elevation above sea level: | 137 m (450 ft) |
| Net sludge yield (kg MLSS/kg $BOD_5$): | 0.76 |
| DO mixed liquor concentration ($C_O$) | 2 mg/L |
| Oxygen coefficients: | |
| kg $O_2$/kg $BOD_5$ | 1.28 |
| kg $O_2$/kg $NH_3$-N | 4.60 |
| Transfer factors: | |
| $\alpha$ (typical for coarse bubble diffuser) | 0.85 |
| $\beta$ (typical for domestic wastewater) | 0.95 |
| Sludge settling zone overflow rate: | < 10 $m^3/m^2/d$ |
| Aeration tank detention time: | 1 day |
| Typical $O_2$ transfer rate for coarse bubble diffusers: | 30 kg$O_2$/kW-d |
| | (48 lb$O_2$/hp-d) |
| Solids retention: | 25 days |

   *a. Sludge production.* Calculate the sludge production rate based on the desired $BOD_5$ removal. Calculate the system total solids mass based on the sludge production rate and the assumed solids retention time, as shown in Table E-2.

   *b. Aeration power.* Calculate the blower capacity based on the sludge production rate, desired TKN removal/synthesis, and the site specific conditions, as shown in Table E-2.

   *c. Unit dimensions.* Estimate the required unit process dimemsions including the chlorine contact tank based on one-day hydraulic detention time and using two sludge settling hoppers, as shown in Table E-2.

**Table E-2**
**Package Plant Extended Aeration Design Calculations**

a. Aerobic Volume

$$BOD_5 \ Removed \ (kg/d) = \frac{Flow \ (L/d)}{10^6 \ (mg/kg)} \times (BOD_{influent} - BOD_{effluent}) \ (mg/L)$$

$$BOD_5 \ Removed = \frac{125 \times 10^3}{10^6} \times (250 - 20) \cong 29 \ kg/d \ (64 \ lb/d)$$

$$Sludge \ Production \ (kg/d) = Net \ Sludge \ Yield \ (kg \ MLSS/kg \ BOD_5) \times BOD_5 \ Removed \ (kg/d)$$

$$Sludge \ Production = 0.76 \times 29 = 22 \ kg/d \ (48.5 \ lb/d)$$

$$System \ Mass \ (kg) = Sludge \ Production \ (kg/d) \times Solids \ Retention \ (d)$$

$$System \ Mass = 22 \ (kg/d) \times 25 \ (d) = 550 \ kg \ (1 \ 213 \ lb)$$

b. Aeration Power

$$NH_3 - N_{oxidized} = TKN_{influent} - Synthesis \ N - TKN_{effluent}$$

$$Synthesis \ N = 5\% \ waste \ activated \ sludge \ of \ total \ daily \ sludge \ production$$

$$Synthesis \ N = 0.05 \times 22 \ (kg/d) = 1.1 \ kg/d \ (2.4 \ lb/d)$$

$$Synthesis \ N \ (mg/L) = \frac{1.1 \ (kg/d) \times 10^6 \ (mg/kg)}{125 \times 10^3 \ (L/d)} = 8.8 \ mg/L$$

$$NH_3 - N_{oxidized} = 40 \ (mg/L) - 8.8 \ (mg/L) - 5 \ (mg/L) = 26.2 \ mg/L$$

$$NH_3 - N \ (kg/d) = 26.2 \ (mg/L) \times 10^{-6} \ (kg/mg) \times 125 \times 10^3 \ (L/d) = 3.28 \ kg/d \ (7.2 \ lb/d)$$

$$AOR = 1.28 \ (kgO_2/kgBOD_5) \times Synthesis \ N \ (kgBOD_5/d) + 4.6 \ (kgO_2/kg \ NH_3-N) \times NH_3-N_{oxidized} \ (kg/d)$$

$$AOR = 1.28 \ (kgO_2/kgBOD_5) \times 1.1 \ (kgBOD_5/d) + 4.6 \ (kgO_2/kg \ NH_3-N) \times 3.28 \ (kgNH_3-N/d)$$

$$AOR = 16.5 \ kgO_2/d \ (36.4 \ lbO_2/d)$$

where:

AOR = Actual Oxygen Requirements (kg $O_2$/d)

$$SAOR = AOR \times \frac{C_S \ (mg/L) \times \Theta^{(20-T)}}{\alpha \times (\beta \times C_{SW} - C_O)}$$

where:

SAOR = Standard Actual Oxygen Requirements (kg $O_2$/d)
$\Theta$ (temperature correction factor) = 1.024

(Sheet 1 of 3)

Table E-2 (Continued)

$C_S$ ($O_2$ saturation concentration at standard temperature and pressure) = 9.02 mg/L
$C_{SW}$ = correction factor for elevation ( i.e., 450 ft) = 9.02 - 0.0003 × elevation
$C_{SW}$ = 9.02 - 0.0003 × 450 = 8.88 mg/L (<u>NOTE:</u> 0.0003 may be used as rule-of-thumb describing a 0.0003 mg/L rise/drop in DO saturation concentration per every foot of elevation increase/decrease.)
$C_O$ = 2 mg/L
$\alpha$ = 0.85; $\beta$ = 0.95; T = 17°C (62°F)

$$SAOR = 16.5\,(kgO_2/d) \times \frac{9.02\ (mg/L) \times 1.024^{(20-17)}}{0.85 \times [0.95 \times 8.88\ (mg/L) - 2.0\ (mg/L)]} = 29.2\ kgO_2/d\ (64.4\ lbO_2/d)$$

$$Motor\ Requirements\ (kW) = \frac{SAOR\ (kgO_2/d)}{O_2\ Transfer\ Rate\ (kgO_2/kW\text{-}d)}$$

$$Motor\ Requirements\ (kW) = \frac{29.2\ (kgO_2/d)}{30\ (kgO_2/kW\text{-}d)} = 1.0\ kW\ (1.3\ hp)$$

Since blowers typically have an efficiency of 50% or less, select 2 aerators with 3.73 kW (5 hp) motors.

c. Unit Dimensions

1. Aeration Tank Volume = 125 m³ at one day hydraulic detention

Assume tank dimensions based on values typical of extended aeration systems:
Operating depth = 3.048 m (10 ft)
Width = 2.895 m (9.5 ft)

$$Tank\ Length = \frac{Volume}{Width \times Depth} = \frac{125}{2.895 \times 3.048} = 14.2\ m\ (47\ ft)$$

2. Sludge settling zone overflow rate.To calculate the sludge settling zone overflow, assume two sludge settling hoppers each with a top dimension of 2.895 m (9.5 ft) and a bottom dimension of 0.304 m (1 ft).

Surface area of sludge settling zone = 2 × 2.895 × 2.895 = 16.76 m² (180 ft²)

$$Overflow\ Rate = \frac{125\ (m^3/d)}{16.76\ (m^2)} = 7.46\ m^3/m^2\text{-}d\ (24.5\ ft^3/ft^2\text{-}d)$$

Assume settling height above hoppers = 0.3 m (1 ft)

Depth of hopper = 3.048 m - 0.3 m = 2.75 m (9 ft)

$$Hopper\ Volume = \frac{1}{3}\,(A_1 + A_2 + \sqrt{A_1 \times A_2}\,) \times depth$$

$$Hopper\ Volume = \frac{1}{3}\,(2.895^2 + 0.304^2 + \sqrt{2.895^2 \times 0.304^2}\,) \times 2.75 = 8.56\ m^3\ (302\ ft^3)$$

Total Hopper Volume = 8.56 m³ x 2 = 17.1 m³ (604 ft³)

Sludge holding tank shall be located at head of tank and shall equal volume of sludge hoppers.

Width = 2.895 m (9.5 ft)
Depth = 3.048 m (10 ft)

$$Holding\ Tank\ Length\ (m) = \frac{Total\ Hopper\ Volume\ (m^3)}{Width\ (m) \times Depth\ (m)}$$

$$Holding\ Tank\ Length = \frac{17.1\ m^3}{2.895\ m \times 3.048\ m} = 1.94\ m\ (6.4\ ft)$$

(Sheet 2 of 3)

Table E-2 (Concluded)

3. If chlorination is used as a disinfectant, the chlorine contact tank shall have a detention time of 75 min; therefore the tank shall have a capacity of 6.5 m³ and the tank dimensions will be:

Width = 2.895 m (9.5 ft)
Length = 3.048 m (10 ft)

$$Chlorine\ Contact\ Tank\ Depth\ (m) = \frac{Volume\ (m^3)}{Width\ (m) \times Depth\ (m)} = \frac{6.5\ m^3}{2.895\ m \times 3.048\ m}$$

Chlorine Contact Tank Depth = 0.74 m (2.4 ft)

(Sheet 3 of 3)

*d. Equipment specifications.* Figure E-1 presents a plan view and a side view of the pre-engineered package plant extended aeration with the following specifications:

(1) The unit package plant will require no pre-treatment other than wastewater pumping from an influent manhole lift station.

(2) The influent pipe shall have a minimum of a 150 mm (6 in.) diameter from the influent manhole and will discharge directly to a combination comminutor/bar screen located ahead of (and on top of) the aeration tank.

(3) Two 3.73 kW (5 hp) blower assemblies shall provide air at 31 kPa (4.5 psi) to ensure a 2.0 mg/L DO level in the aeration tank at all times.

(4) A minimum of 44 diffusers will be required to distribute aeration at the aeration tank floor level. At least six (6) diffusers will be provided in the sludge holding tank and one (1) in the chlorine contact tank.

(5) A totalizing flow meter will be provided to record the daily flow patterns and total.

(6) A minimum of eight spray nozzles will be required on the top of the aeration tank on the side opposite to the aeration diffuser drops.

(7) Each sludge hopper will be equipped with an air lift pump with openings 150 mm (6 in.) above the hopper bottoms.

(8) The air lift pumps will discharge to a combination 75 mm (3 in.) sludge return and sludge waste line to the head of the tank.

(9) Blower units shall be controlled by a blower panel located above the aeration tank.

(10) Scum skimmers will be provided at a scum baffle ahead of the tank discharge (by V-notch weir) to the chlorine contact tank.

(11) Should ultraviolet disinfection be chosen in lieu of chlorination of tank effluent, an in-pipe rather than open channel effluent flow may be specified.

E-4

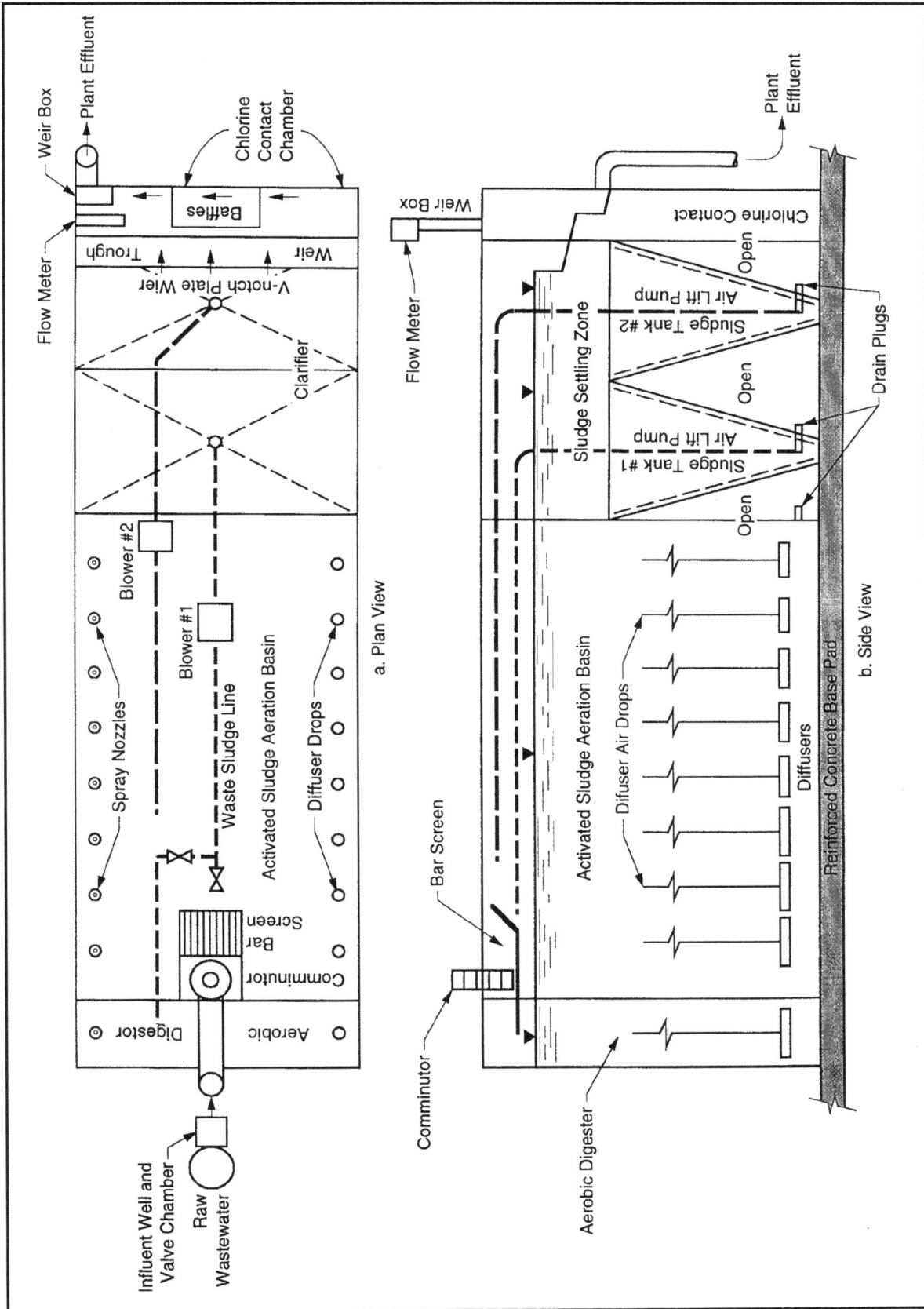

Figure E-1. Pre-engineered package plant extended aeration

## E-2. Oxidation Ditch (Continuous-Loop Reactor) Carrousel—Wraparound

Design a carrousel (circular or wraparound) oxidation ditch to treat municipal wastewater at an average influent flow rate of 378 500 L/d (100,000 gal/d). The new system will use mechanical aerators and have the design parameters shown in Table E-3.

**Table E-3**
**Design Parameters and Assumptions**

*Influent/Effluent Composition*

| Parameter | Influent | Effluent |
|---|---|---|
| BOD$_5$ | 250 mg/L | 5 mg/L |
| TSS | 300 mg/L | 10 mg/L |
| TKN | 30 mg/L | 5 mg/L |
| NH$_3$-N | -- | 0.5 mg/L |

*Assumptions*

| | |
|---|---|
| Minimum wastewater temperature: | 16°C (61°F) |
| Process solids retention time: | 20 days |
| MLSS concentration: | 4000 mg/L |
| Net yield (kg MLSS/ kg BOD$_5$): | 0.76 |
| Oxygen coefficients: | |
|   kg O$_2$/kg BOD$_5$ | 1.28 |
|   kg O$_2$/kg NH$_3$-N | 4.60 |
| Transfer factors: | |
|   α (typical for mechanical aerator) | 0.90 |
|   β (typical for domestic wastewater) | 0.95 |
| Typical O$_2$ transfer rate for mechanical aerator: | |
| | 37 kgO$_2$/kW-d |
| Site elevation (sea level + tank height): | (60 lbO$_2$/hp-d) |
| Clarifier overflow rate: | 9.1 m (30 ft) |
| Side water depths, | 16.3 m³/m²/d |
| clarifier and reactor channels: | 3.04 m (10 ft) |

    *a.  Carrousel volume.*  Calculate the sludge production rate based on the desired BOD$_5$ removal. Calculate the system total solids mass based on the sludge production rate and the assumed solids retention time. Calculate the carrousel volume from the calculated system total solids mass and the assumed MLSS concentration, as shown in Table E-4.

    *b.  Aeration power.*  Calculate the blower capacity based on the desired TKN removal/synthesis and the site specific conditions, as shown in Table E-4.

    *c.  Clarifier diameter.*  Estimate the required wraparound clarifier diameter based on the assumed clarfier overflow rate and the side water depths, as shown in Table E-4.

    *d.  Carrousel specifications.*  The carrousel shown in Figure E-2 has the following specifications:

| | |
|---|---|
| Clarifier diameter: | 6.4 m (21 ft) |
| Inner channel: | 4 m (13 ft) |
| Outer channel: | 4 m (13 ft) |
| Entire tank diameter: | 14 m (46 ft) |
| Walls and miscellaneous equipment thickness: | 2 m (6.5 ft) |
| Constructed carrousel diameter: | 14 m + 2 m = 16 m (52.5 ft) |

**Table E-4**
**Oxidation Ditch Design Calculations**

a. Carrousel Volume

$$BOD_5 \; Removed \; (kg/d) = \frac{Flow \; (L/d)}{10^6 \; (mg/kg)} \times (BOD_{influent} - BOD_{effluent}) \; (mg/L)$$

$$BOD_5 \; Removed = \frac{378.5 \times 10^3}{10^6} \times (250 - 5) = 92.7 \; kg/d \; (204 \; lb/d)$$

$$Sludge \; Production \; (kg/d) = Net \; Yield \; (kg \; MLSS/kg \; BOD_5) \times BOD_5 \; Removed \; (kg/d)$$

$$Sludge \; Production = 0.76 \times 92.7 = 70.5 \; kg/d \; (156 \; lb/d)$$

$$System \; Mass = 70.5 \; (kg/d) \times 20 \; (d) = 1 \; 410 \; kg \; (3 \; 107 \; lb)$$

$$Carrousel \; Volume \; (m^3) = \frac{System \; Mass \; (kg) \times 10^3}{MLSS \; Concentration \; (mg/L)}$$

$$Carrousel \; Volume = \frac{1.41 \times 10^3 \times 10^3}{4 \times 10^3} = 353 \; m^3 \; (12{,}466 \; ft^3)$$

b. Aeration Power

*Synthesis N = 5% wasted activated sludge of total daily sludge production*

*Synthesis N = 0.05 × 70.5 (kg/d) = 3.52 kg/d (7.75 lb/d)*

*Synthethis N = 9.3 mg/L (for a daily flow of 378 500 L/d)*

$$NH_3 - N_{oxidized} = TKN \; (mg/L) - Synthesis \; N \; (mg/L) - N\text{-}NH_{3(effluent)} \; (mg/L)$$

$$NH_3 - N_{oxidized} = 30 - 9.3 - 0.5 = 20.2 \; mg/L$$

$$NH_3 - N_{oxidized} = 7.6 \; kg/d \; (16.9 \; lb/d) \text{ - based on a daily flow of 378 500 L/d}$$

$$AOR = kg \; O_2/kg \; BOD_5 \times Synthesis \; N \; (kg \; BOD_5/d) + kg \; O_2/kg \; NH_3\text{-}N \times NH_3\text{-}N_{oxidized} \; (kg/d)$$

$$AOR = 1.28 \times 3.52 + 4.6 \times 7.6 = 4.5 + 35.0 = 39.5 \; kg \; (87 \; lb)$$

where:
  AOR = Actual Oxygen Requirements (kg $O_2$/d)

$$SAOR = AOR \times \frac{C_S \; (mg/L) \times \Theta^{(20 - T)}}{\alpha \times (\beta \times C_{SW} - C_O)}$$

$$Clarifier \; Area \; (m^2) = \frac{Design \; Flow \; (m^3/d)}{Overflow \; Rate \; (m^3/m^2/d)}$$

where:

  SAOR = Standard Actual Oxygen Requirements (kg$O_2$/d)
  $\Theta$ (temperature correction factor) = 1.024
  $C_S$ (DO saturation concentration at standard temperature and pressure conditions) = 9.02 mg/L
  $C_{SW}$ = Correction factor for elevation (i.e., 30 ft) = 9.02 - 0.0003 × elevation
  $C_{SW}$ = 9.02 - 0.0003 × 30 = 9.011 mg/L (NOTE: 0.0003 may be used as rule-of-thumb describing a 0.0003 mg/L rise/drop in DO saturation concentration per every foot of elevation increase/decrease.)

(Sheet 1 of 3)

Table E-4 (Continued)

$C_O$ = 2.0 mg/L
$\alpha$ = 0.90; $\beta$ = 0.95; T = 16°C (61°F)

$$SAOR = 39.5\ (kgO_2/d) \times \frac{9.02\ (mg/L) \times 1.024^{(20-16)}}{0.90 \times [0.95 \times 9.01\ (mg/L) - 2.0\ (mg/L)]}$$

$$= 66.4\ kg\ O_2/d\ (146\ lb\ O_2/d)$$

$$Aerator\ Power\ Requirements\ (kW) = \frac{SAOR\ (kgO_2/d)}{O_2\ Transfer\ Rate\ (kgO_2/kW\text{-}d)}$$

$$Aerator\ Power\ Requirements = \frac{66.4\ (kgO_2/d)}{37\ (kgO_2/kW\text{-}d)} = 1.80\ kW\ (2.4\ hp)$$

Since blowers typically have an efficiency of 50% or less, select two aerators with 3.73 kW (5 hp) motors or two 2-speed aerators with 7.5/5 hp motors per basin normally operated on low speed.

c. Clarifier Diameter

Design flow = 387 500 L/d

Overflow rate = 16.3 m³/m²/d

$$Clarifier\ Area\ (m^2) = \frac{Design\ Flow\ (m^3/d)}{Overflow\ Rate\ (m^3/m^2/d)}$$

$$Clarifier\ Area = \frac{378.5\ (m^3/d)}{16.3\ (m^3/m^2/d)} = 23.2\ m^2\ (250\ ft^2)$$

$$Clarifier\ Surface\ Area = \frac{\pi D^2}{4} = 23.2\ m^2$$

Solve for diameter (D):

$$D = \sqrt{23.2/0.785} \approx 6.0\ m\ (20\ ft)$$

A new clarifier surface area is then recalculated:

$$New\ Clarifier\ Area\ (m^2) = \frac{\pi \times D^2}{4}$$

$$New\ Clarifier\ Area = \frac{\pi \times [6\ (m)]^2}{4} = 28.3\ m^2\ (305\ ft^2)$$

Detention time (hr) = Volume (m³) × 24 (hr/d)/Flow (m³/d)

Volume = Area (m²) × Depth (m) = 28.3 (m²) × 3.04 (m) = 86 m³ (3 038 ft³)

(Sheet 2 of 3)

**Table E-4  (Concluded)**

$$\text{Detention time} = \frac{86 \ (m^3) \times 24 \ (hr/day)}{378.5 \ (m^3/d)} = 5.5 \ hr$$

*Total clarifier diameter = water diameter + 2 × wall thickness*

*Total clarifier diameter = 6 m + 2 × 0.2 m = 6.4 m (21 ft)*

For a 3.04 m side water depth (SWD) for the channels

$$\text{Volume} = 378.5 \ m^3 \ of \ wraparound \ channels$$

$$A = \frac{\pi D_1^2}{4} - \frac{\pi D_2^2}{4} = 378.5 \ m^3/SWD$$

$$0.785 \ D_1^2 - 0.785 \ D_2^2 = 378.5/3.04$$

$$0.785 \ D_1^2 - 0.785 \times 6.4^2 = 124.5$$

$$D_1 = 14 \ m \ (46 \ ft)$$

(Sheet 3 of 3)

## E-3.  Stabilization Pond

Design a facultative stabilization pond with primary treatment (clarifier and anaerobic digester of Imhoff tank design) to be followed by secondary clarification to treat a domestic wastewater flow of 378 500 L/d (100,000 gal/d).  Influent $BOD_5$ will be 250 mg/L.  Assume a primary clarifier removes 33 percent of the influent $BOD_5$ ($BOD_5 = 0.68 \ BOD_u$), and influent wastewater $[SO_4^{2-}]$ is $\leq 500$ mg/L.  Four rectangular ponds in parallel are to be constructed.  The controlling winter temperature of each pond will be 4.5 °C (40 °F).  Length to width ratio of each pond will be 3:1, as is typical for such facilities.

Find, as shown in Table E-5:

Total area of ponds.

Applied $BOD_5$ loading.

Dimensions of the ponds.

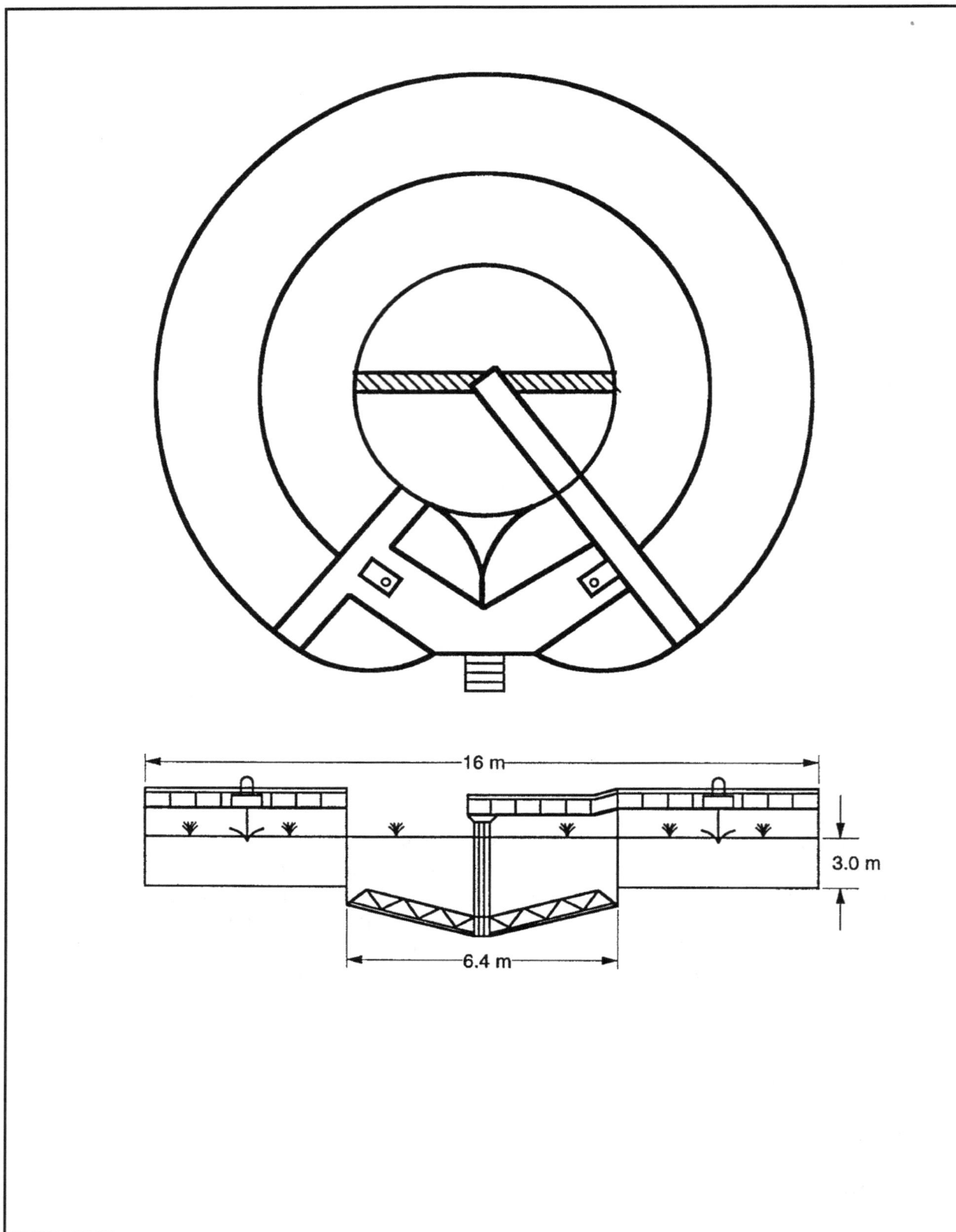

Figure E-2. Oxidation ditch carrousel wraparound (closed-loop reactor)

EM 1110-2-501
1 Feb 99

E-10

**Table E-5**
**Stabilization Pond Design Calculations**

$BOD_5$ (primary clarifier effluent) = 250 (mg/L) × (1-0.33) = 167.5 mg/L

$S_i = BOD_u = 167.5/0.68 = 246.3$ mg/L

$t = \dfrac{V}{Q} = 0.035 \times 246.3 \times [1.085^{(35-4.5)}] \times 1 \times 1 = 104$ days

$V$ (4 basins) $= Q$ (m$^3$/d) $\times t$ (d) $= 378.5$ (m$^3$/d) $\times 104$ (d) $= 39\ 364$ m$^3$ (31.9 acre-ft)

$A$ (4 basins) $= \dfrac{V}{d}$

where:

A = total ponds area (hectares)
d = facultative pond effective depth = 1.8 m (6 ft); to include 0.3 m (1 ft) for sludge storage

$A$ (4 basins) $= \dfrac{39\ 364\ (m^3)}{1.8\ (m)} \times \dfrac{1\ ha}{10\ 000\ (m^2)} = 2.19$ ha (5.41 ac)

$BOD_5$ Load $= 387.5 \times 10^3$ (L/d) $\times 167.5$ (mg/L) $\times 10^{-6}$ (kg/mg) $= 64.9$ kg/d (142.9 lb/d)

Applied Load $= 64.9$ (kg/d) $\times \dfrac{1}{2.19\ ha}$

Applied Load = 29.6 kg BOD$_5$/ha-d (26.5 lb BOD$_5$/acre-d)

Each Pond Area $= \dfrac{Total\ Pond\ Area\ (ha) \times 10\ 000\ (m^2/ha)}{4} = 5\ 475$ m$^2$ (17 958 ft$^2$)

Pond Area = Length ×Width

Length = 3 × width (W)

Pond Area = (3W) × W = 5 475 m$^2$

Width (W) = 42.7 m (140 ft)

Length = 128 m (420 ft)

Use the Gloyna equation ("Facultative Waste Stabilization Pond Design in Ponds as a Wastewater Treatment Alternative," by E. F. Gloyna, J. F. Malina, Jr. and E. M. Davis):

$$t = \frac{V}{Q} = CS_i\ [\theta^{(35-T)}]ff'$$

where

$V$ = pond volume, m$^3$.

$C$ = conversion coefficient = 0.035 (a constant metric conversion).

$Q$ = flow (378.5 m$^3$/d).

$S_i$ = ultimate influent $BOD_5$ mg/L.

$f$ = sulfide or immediate chemical oxygen demand = 1 (for[ $SO_4^{2-}$] concentrations $\leq 500$ mg/L).

$f$ = algae toxicity factor = 1.

$\theta$ = temperature coefficient (the value of $\theta$ ranges from 1.036 to 1.085, and 1.085 is recommended as it is conservative).

$T$ = average water temperature for the pond during winter months, °C.

$t$ = hydraulic detention time (days).

## E-4. Zero Discharge or Water Recycle/Reuse For Toilet Flush Water in Rest Areas (Closed-Loop Reuse)

*a.  Background.*

(1)  A combination of an extended aeration-activated sludge wastewater treatment system followed by mixed-media pressure filtration has been successful in treating liquid waste from a comfort facility with eight water closets and two urinals, plus lavatories. The design wastewater flow is 37 800 L/d (10,000 gal/d).

(2)  The closed-loop reuse principle is generally applicable where liquid discharges from a recreational area are not permitted or desired. After the system is initially filled and operational, a fraction of the treated wastewater (about 6 percent) is fed to the terminal holding pond or lagoon to evaporate to account for the makeup water used for lavoratories and drinking fountains. The makeup water is estimated to represent about 6 percent of total water use. The sludge from the waste solids holding basin is periodically removed by tank truck. The design parameters for the original extended aeration treatment system are presented in Table E-6. Figure E-3 presents a schematic flow diagram of the wastewater recycle-reuse system.

(3)  It is to be expected that 90 to 95 percent of water used in a comfort station facility is for the water closets or toilet flushing functions. Generally, 10 to 20 cycles are required for the system to reach equilibrium with an input of 5 to 10 percent of potable water for the lavatories or drinking fountains. The wastewater from lavatories and drinking fountains is considered "new" water and is a factor in the control of the amount of wastewater that must be fed to the final holding pond to evaporate.

(4)  Operating records reveal no objectionable odors from the water closets or lavatories, no objectionable colors from blue (or other food dyes) introduced to give a sanitized look to the flushing waters, no foaming in the sanitary facilities, and no building of total suspended solids. The 90 to 95 percent of reused water in the water closets and urinals has an acceptable quality following the extended aeration process and multimedia filtration.

(5)  Use surveys to indicate that toilet flush water use is about 12.7 L (3.4 gal) per flush and 15.0 L (4.0 gal) per toilet user. Potable water use (lavatories and drinking fountains) is approximately 0.8 L (0.2 gal) per toilet user. Average resident time in the toilet facility is expected to be 3 min.

*b.  Recycled wastewater.*

(1)  The desired treatment characteristics of the recycled wastewater are shown in Table E-7.

**Table E-6**
**Original Design Parameters of the Existing Extended-Aeration System**

| Parameter | Value |
|---|---|
| Design flow: | 37 800 L/d (10,000 gal/d) |
| Aeration Tank: | |
|    Detention time | 24 hr |
|    Volume | 38 m$^3$ (1342 ft$^3$) |
|    Oxygen transfer rate | 756 g/hr (1.7 lb/hr) |
| Max. Return Solids flow: | 1.83 L/s (0.48 gal/s) |
| Waste solids holding basin: | 15.1 m$^3$ (533 ft$^3$) |
| Comminutor: | 8.8 L/s (2.3 gal/s) |
| Settling basin: | |
|    Detention time | 4 hr |
|    Volume | 6.3 m$^3$ (222 ft$^3$) |
| Holding pond: | |
|    Volume | 567 m$^3$ (20,000 ft$^3$) |
|    Surface area | 497 m$^2$ (5350 ft$^2$) |

The comminutor shreds to 6 mm (0.2 in). An overflow bypass around the comminutor has a manually cleaned medium bar screen. All pumps are pneumatic.

**Table E-7**
**Desired Characteristics of Recycled Wastewater**

| Parameter | Concentration Range |
|---|---|
| MLSS | 3000 to 5000 mg/L |
| Settleability | 200 to 850 mL |
| (as determined by the MLSS volume in 1 L graduated cylinder after 1 hr) | |
| Alkalinity | |
| pH | |
| TSS | 50 to 500 mg/L |
| | 5.5 to 8.3 |
| | < 15 mg/L |

(2) To achieve the best operation, the recycled wastewater must be chemically stable and the total suspended solids and total volatile solids must remain relatively constant. The most desirable range for MLSS would probably be 3500 to 4000 mg/L with an accompanying settleability of 400 to 600 mL.

   *c.   Unit processes for closed-loop reuse.*

(1) The unit processes shown in Table E-8 have been added for the closed-loop reuse to meet the desired characteristics identified in Table E-7.

(2) The multimedia rapid filtration pressurized vessel has a design filtration rate of 80 to 160 L/min/m$^2$ and a backwash design flow rate of 285 to 610 L/min/m$^2$. The filter appears to operate best at a filtration rate of 94 L/min/m$^2$ (2.3 gal/min/ft$^2$) and at a backwash cleaning rate of 345 L/min/m$^2$ (8.5 gal/min/ft$^2$). Total suspended solids in the recycled wastewater must be less than 15 mg/L for reuse in the toilet facility.

## E-5. Sequencing Batch Reactor (SBR)

   *a.   General.*

(1) The design of a sequencing batch reactor (SBR) involves the same factors commonly used for the flow-through activated sludge system. The aspects of a municipally treated waste which require

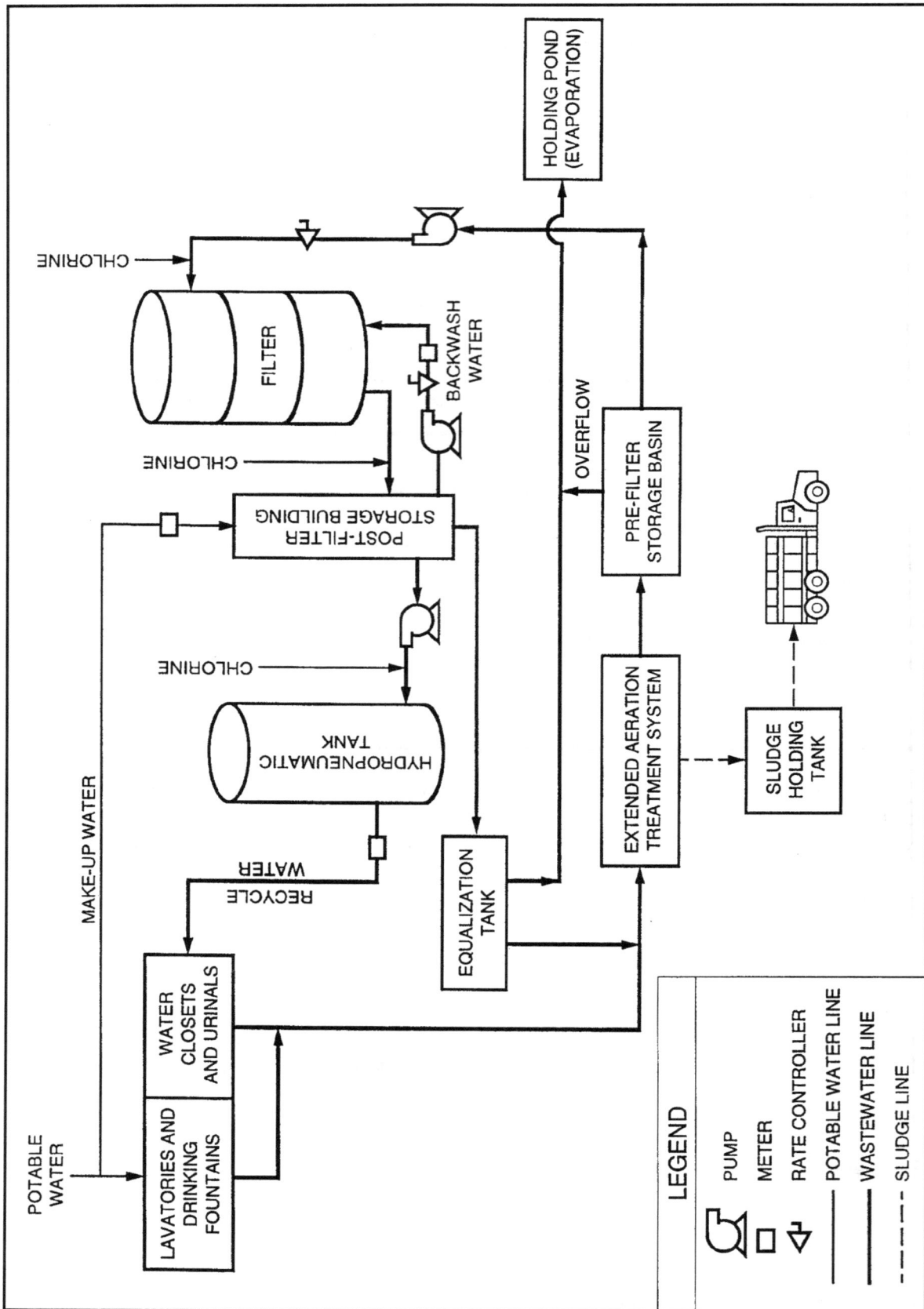

Figure E-3.  Flow diagram for wastewater recycle-reuse system

Table E-8
Unit Processes for the Closed Loop System

| Unit Process | Design Parameter |
|---|---|
| Pressure Filter | |
| Diameter | 1.8 m (6 ft) |
| Media | Granular nonhydrous aluminum silicate |
| | (Effective size = 0.57, |
| | Uniform coefficient = 1.66) |
| | 4.1 L/s (65 gal/min) |
| Max. pump rate | 1.6 L/s/m² (0.04 gal/s/ft²) |
| Filtration rate | 2.65 L/s/m² (0.06 gal/s/ft²) |
| Surface wash rate | 5.77 L/s/m² (0.13 gal/s/ft²) |
| Backwash rate | |
| Pre-Filter Storage Tank | 75.6 m³ (2670 ft³) |
| Post-Filter Storage Tank | 75.6 m³ (2670 ft³) |
| Equalization Tank | 18.9 m³ (668 ft³) |
| Hydropneumatic Tank | |
| Total volume | 18.9 m³ (668 ft³) |
| Operating volume | 5.3 m³ (187 ft³) |

dentrification as well as nitrification plus biological phosphorous removal need additional design considerations. Pretreatment of the wastewater before influent in the SBR reactor system is also required.

(2) The following example should be considered an outline to identify reactor volume elements, a diffused aeration system, the basis for signing effluent decanter units, and waste sludge systems for a system receiving 378 500 L/d (100,000 gal/d) of wastewater.

(3) Food-to-mass (F/M) ratio typically ranges from 0.05 to 0.30 with domestic waste F/M ratios typically ranging from 0.10 to 0.15. At the end of the decant phase, the MLSS concentration may vary between 2000 and 5000 mg/L. A typical value for a municipal waste would be 3500 mg/L. The MLSS concentration changes continuously throughout an SBR operating cycle from a maximum at the beginning of a fill phase to a minimum at the end of the react phase.

   b.   *Reactor volume.* Calculate the reactor volume based on the desired $BOD_5$ removal, the F/M ratio, and the MLSS. The F/M ratio and the MLSS at the low water level determine the reactor volume at the low water level, as shown in Table E-10.

   c.   *Decant volume.* Calculate the decant volume as the difference between the reactor volume and the low water volume, as shown in Table E-10. Each operating cycle is normally composed of *mixed fill*, *react fill*, *settle*, *decant*, *sludge waste*, and *idle*. The number of cycles dictates the number of decants per day or the volume of liquid to be decanted for each cycle. The volume per decant per cycle must be selected based on the maximum sustained daily flow.

   d.   *Detention time.* Calculate the maximum detention time based on the reactor volume. Calculate the minimum detention time based on the decant volume, as shown in Table E-10.

   e.   *SBR dimensions.* Estimate the required unit process dimensions, as shown in Table E-10. The basin length can be estimated based on a recommended minimum depth. The minimum depth after decant is determined as the depth of a clarifier in a flow-through system, i.e., quiescent settling and a large settling area. A minimum depth of 2.75 m (9 ft) is typically recommended by designers.

**Table E-9**
**Design Assumptions**

*Influent/Effluent Composition*

| Parameter | Influent | Effluent |
|---|---|---|
| BOD5 | 250 mg/L | 25 mg/L |
| TSS | 250 mg/L | 30 mg/L |
| NH3-N | 25 mg/L | 1 mg/L |
| Total Phosphorous | 10 mg/L | 2 mg/L |
| TKN | 40 mg/L | 5 mg/L |

*Assumptions*

| | |
|---|---|
| F/M Ratio (kgBOD$_5$applied/kgMLSS-d): | 0.13 |
| MLSS: | 3500 mg/L |
| Minimum clarifier depth: | 2.75 m (9 ft) |
| Net sludge yield (kg MLSS/kg BOD$_5$): | |
| Wastewater temperature: | 0.76 |
| Minimum solids retention time: | 20°C(67°F) |
| Reactor volume decanted each day: | 8 d |
| Net elevation above sea level: | 60% |
| DO mixed liquor concentration (C$_o$): | 304 m (1000 ft) |
| Oxygen coefficients: | 2 mg/L |
|    kg O$_2$/kg BOD$_5$ | |
|    kg O$_2$/kg NH$_3$-N | 1.28 |
| Transfer factors: | 4.60 |
|    α (typical for coarse bubble diffuser) | |
|    β (tyical for domestic wastewater) | 0.85 |
| Typical O2 transfer rate for coarse bubble diffusers: | |
| | 0.95 |
| Number of cycles per day: | |
| Include two square basins for operational flexibility | 1.25 kgO$_2$/kW-hr (2 lbO$_2$/hp-hr) 4 |

*f.  Aeration power.* Calculate the blower capacity based on the sludge production rate and the total oxygen demand, as shown in Table E-10. In sizing the aeration equipment, it must be noted that the equipment operates only a portion of the SBR operating cycle (part of the fill plus react phases). Therefore, the total daily oxygen requirements must be met in a shorter time period than in a conventional activated sludge flow-through system. The total daily oxygen requirements are estimated by adding the carbonaceous oxygen demand (oxygen required for BOD$_5$ oxidation) to the nitrogeneous oxygen demand (oxygen required for TKN oxidation).

*g.  Sludge and decanter flows.* Calculate the sludge and decanter flow rates at design conditions, as shown in Table E-10.

*h.  Equipment specifications.* Figure E-4 presents a plan view of the SBR system with the following specifications:

(1)  At least two basins are provided in an SBR design to provide operational flexibility and improved effluent quality. SBR unit dimensions:

> maximum volume = 470 m$^3$ (16,598 ft$^3$).
> high water level = 5.80 m (19 ft).
> maximum decant height = 3.05 m (10 ft).
> low water level = 2.75 m (9 ft).
> hydraulic detention time at low water level = 17.9 hrs.

**Table E-10**
**SBR Design Calculations**

a. *Reactor Volume*

$$BOD_5 \text{ Removed } (kg/d) = [(BOD_{influent} - BOD_{effluent}) \ (mg/L)] \times Flow \ (L/d) \times 10^{-6} \ (kg/mg)$$

$$BOD_5 \text{ Removed } = (250 - 25) \times 378.5 \times 10^{-3} = 85.2 \ kg/d \ (187 \ lb/d)$$

$$Required \ Aerobic \ Mass \ (kg) = \frac{BOD_5 \ Removed \ (kg/d)}{F/M \ Ratio \ (kgBOD_5/kbMLSS-d))}$$

$$Required \ Aerobic \ Mass = \frac{85.2 \ (kgBOD_5/d)}{0.13 \ (kgBOD_5/kgMLSS-d))} = 656 \ kg \ MLSS$$

$$Reactor \ Volume_{(low \ water \ level)} \ (m^3) = \frac{MLSS \ Mass \ (kg)}{MLSS \ Concentration \ (mg/L)} \times \frac{10^6 \ (mg/kg)}{10^3 \ (L/m^3)}$$

$$Reactor \ Volume_{(low \ water \ level)} = \frac{656 \ (kg)}{3\ 500 \ (mg/L)} \times 10^3 \ (mg-m^3/kg-L) = 188 \ m^3 \ (6\ 634 \ ft^3)$$

Since the decant volume represents 60% of the total volume,

Total Reactor Volume = 188/(1-0.6) = 470 $m^3$ (16 598 $ft^3$)

b. *Decant Volume*

Total Decant Volume = Total Reactor Volume ($m^3$) - Reactor Volume$_{(low \ water \ level)}$ ($m^3$)

Total Decant Volume = 470 ($m^3$) - 188 ($m^3$) = 282 $m^3$ (9 959 $ft^3$)

c. *Detention Time*

$$Max. \ Detention \ Time \ (hr) = \frac{Total \ Reactor \ Volume \ (m^3)}{Flow \ (L/d) \times 10^{-3} \ (m^3/L)} \times 24 \ (hr/d)$$

$$Max. \ Detention \ Time = \frac{470 \ (m^3) \times 24 \ (hr/d)}{378.5 \times 10^3 \ (L/d) \times 10^{-3} \ (m^3/L)} = 30 \ hr$$

$$Min. \ Detention \ Time \ (hr) = \frac{Decant \ Volume \ (m^3)}{Flow \ (L/d) \times 10^{-3} \ (m^3/L)} \times 24 \ (hr/d)$$

$$Min. \ Detention \ Time = \frac{282 \ (m^3) \times 24 \ (hr/d)}{378.5 \times 10^3 \ (L/d) \times 10^{-3} \ (m^3/L)} = 17.9 \ hr$$

d. *SBR Dimensions*

$$Basin \ Area \ (m^2) = \frac{Basin \ Volume_{low \ water \ level} \ (m^3)}{Minimum \ Depth \ (m)} = \frac{188 \ (m^3)}{2.75 \ (m)} = 68.4 \ m^2 \ (736 \ ft^2)$$

$$Basin \ Length = \sqrt{68.4 \ (m^2)} \cong 9 \ m \ (30 \ ft)$$

(Sheet 1 of 3)

Table E-10. (Continued)

$$Basin\ Depth\ (m) = \frac{Total\ Reactor\ Volume\ (m^3)}{Basin\ Area\ (m^2)} = \frac{470\ (m^3)}{[9\ (m)]^2} = 5.8\ m\ (19\ ft)$$

e. Aeration Power

$$Nitrogeneous\ O_2\ Demand\ (kg\ O_2/d) = NH_3\text{-}N_{oxidized}\ (kg/d) \times kg\ O_2/kg\ BOD_5$$

$$NH_3\text{-}N_{oxidized}\ (kg/d) = TKN\ Removed\ (kg/d) - Synthesis\ N\ (kg/d)$$

$$TKN\ Removed = (40 - 5) \times 378.5 \times 10^{-3} = 13.25\ kg/d\ (29\ lb/d)$$

Synthethis N = 5% waste activated sludge of total daily sludge production

$$Sludge\ Production\ (kg/d) = Net\ Sludge\ Yield\ (kgMLSS/kgBOD_5) \times BOD_5\ Removed\ (kg/d)$$

$$Sludge\ Production = 0.76\ (kgMLSS/kgBOD_5) \times 85.2\ (kg/d) = 64.8\ kg/d\ (143\ lb/d)$$

$$Synthethis\ N = 0.05 \times 64.8\ (kg/d) = 3.24\ kg/d\ (7\ lb/d)$$

$$NH_3\text{-}N_{oxidized} = 13.25\ (kg/d) - 3.24\ (kg/d) = 10\ kg/d\ (22\ lb/d)$$

$$Nitrogeneous\ O_2\ Demand = 10\ (kgNH_3\text{-}N_{oxidized}/d) \times 4.6\ (kgO_2/kgNH_3\text{-}N_{oxidized}) = 46\ kgO_2/d$$

$$Carbonaceous\ O_2\ Demand\ (kg\ O_2/d) = BOD_5\ Mass\ (kg/d) \times kg\ O_2/kg\ BOD_5$$

$$Carbonaceous\ O_2\ Demand = 3.24\ (kgBOD_5/d) \times 1.28\ (kgO_2/kgBOD_5) = 4.15\ kgO_2/d$$

AOR (kg/d) = Carbonaceous O2 Demand (kg/d) + Nitrogeneous O2 Demand (kg/d)

$AOR = 4.15\ kgO_2/d + 46\ kgO_2/d = 50.15\ kgO_2/d$

where:

AOR = Actual Oxygen Requirements (kg $O_2$/d)

$$SAOR\ (kg\ O_2/hr) = \frac{AOR \times C_S \times \Theta^{(T-20)}}{\alpha \times (\beta \times C_{SW} - C_O) \times Blower\ Usage\ (hr/d)}$$

where:

SAOR = Standard Actual Oxygen Requirements (kg $O_2$/d)
$\Theta$ (temperature correction factor) = 1.024
$C_S$ ($O_2$ saturation concentration at standard temperature and pressure) = 9.02 mg/L
$C_{SW}$ = concentration correction for elevation (i.e., 1 000 ft) = 9.02 - 0.0003 × elevation
$C_{SW}$ = 9.02 - 0.0003 × 1 000 = 8.72 mg/L (NOTE: 0.0003 may be used as a rule-of-thumb describing a 0.0003 mg/L rise/drop in DO saturation concentration per every foot of elevation increase/decrease.)
$C_O$ = 2 mg/L
$\alpha$ = 0.85; $\beta$ = 0.95; T = 20°C (67°F)
Blower Usage = 14 hr/d (based on 4 cycles per day (6 hr/cycle), 1.0 hr fill time, 3.5 hr react time, 0.75 hr settle time, 0.5 hr decant time, and 0.25 hr idle time)

(Sheet 2 of 3)

Table E-10.  (Continued)

$$SAOR = \frac{50.15 \ (kgO_2/d) \times 9.02 \ (mg/L) \times 1.024^{(20 - 20)}}{0.85 \times (0.95 \times 8.72 - 2) \ (mg/L) \times 14 \ (hr/d)} = 6.1 \ kgO_2/hr \ (13.4 \ lbO_2/hr)$$

$$Motor \ Requirements \ (kW) = \frac{SAOR \ (kgO_2/d)}{O_2 \ Transfer \ Rate \ (kg/kW{-}d)}$$

$$Motor \ Requirements = \frac{6.1 \ (kgO_2/hr)}{1.25 \ (kgO_2/kW{-}hr)} = 4.9 \ kW \ (6.5 \ hp)$$

Since blowers typically have an efficiency of 50% or less, select two aerators with 11.2 kW (15 hp) motors.  Blower size depends on the standard air flow rate.  The standard air flow rate in standard cubic meters per minute (SCMM) is calculated as follows:

$$SCMM = \frac{\dfrac{SAOR \ (kgO_2/d)}{Blower \ Usage \ (hr/d)} \times \dfrac{1 \ hr}{60 \ min}}{O_2 \ Content \ (kgO_2/m^3 \ air) \times Absorption \ Efficiency}$$

where:
   Air $O_2$ Content (at standard conditions) = 0.2793 kg $O_2/m^3$ of air
   Obtain blower absorption efficiency from manufacturers

f.  Sludge and Decant Flows

$$Sludge \ flow \ rate \ (L/d) = \frac{Sludge \ Mass \ Flow \ (kg/d)}{Sludge \ Density \ (kg/L)}$$

Typical sludge density = 1.02 kg/L

$$Decanter \ Flow \ Rate \ (L/min) = \frac{MDF}{NB \times NCB \times MCT}$$

where:
   MDF = maximum daily flow for decant (or sludge waste)
   NB  = number of basins
   NCB = number of cycles per basin
   MCT = maximum cycle time for de cant or sludge waste (min)

(Sheet 3 of 3)

(2)  Blowers: rotary positive displacement.

(3)  Diffusers: 4-10 tube coarse bubble retriever diffuser assembly (2 per basin).

(4)  Mixers: 2 at 3.73 kW (5 hp).

(5)  Sludge pumps: 2 at 1.49 kW (2 hp).

(6)  Decanter sizing:        cycles per day = 4.
                                 volume per decant = 70.5 $m^3$ (2490 $ft^3$).
                                 decant time = 30 min.
                                 decant flow rate = 2.35 $m^3$/min (621 gal/min).

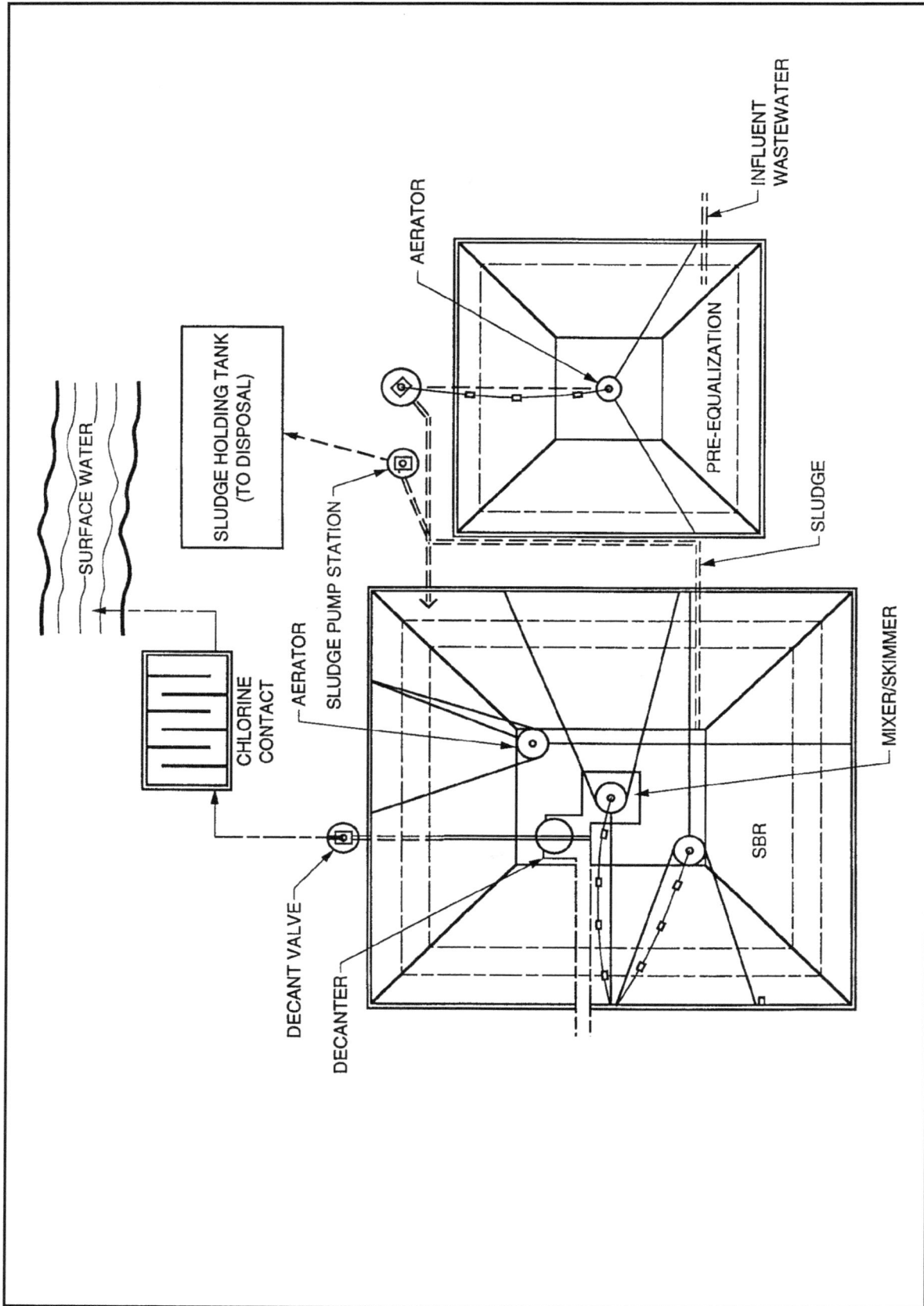

Figure E-4. Sequencing batch reactor

**Table E-11**
**Design Assumptions**

**Influent/Effluent Composition**

| Parameter | Influent | Effluent |
|---|---|---|
| BOD$_5$ | 250 mg/L | < 30 mg/L |
| TSS | 250 mg/L | < 30 mg/L |
| TKN | 20 mg/L | < 5 mg/L |
| Total Phosphorous | 10 mg/L | |
| Winter Temperature | 20°C (67°F) | |

**Assumptions**

| | |
|---|---|
| L:W Ratio (for all basins): | 3:1 |
| Hydraulic Loading: | 200-600 m³/ha-d |
| | (20,000-65,000 gal/acre-d) |
| BOD$_5$ Loading: | 40-80 kg/ha-d |
| | (36-71 lb/acre-d) |
| TKN Reduction: | 70-90% |

(7) Influent valves: 2, each 150 mm (6 in.) diameter.

(8) Air blower values: 2, each 150 mm (6 in.) diameter.

## E-6. Constructed Wetlands Aerobic Non-Aerated Hyacinth System

Design an aerobic non-aerated hyacinth constructed wetlands secondary treatment system for a municipal wastewater flow of 284 000 L/d (75,000 gal/d). The system will require preliminary or pretreatment (Imhoff tank). Disinfection of effluent may be required depending upon regulatory restrictions.

*a. BOD$_5$ loading.* Calculate the influent BOD$_5$ loading using the influent BOD$_5$ concentration and the design flow, as shown in Table E-12.

*b. Basin surface area.* Calculate the required basin surface area at moderate BOD$_5$ loading rate of 50 kg/ha-d and the required area for the primary or first cells at a BOD5 loading rate of 100 kg/ha-d, as shown in Table E-12.

*c. Primary cell dimensions.* Use two primary cells. Calculate the dimensions of the two primary cells, as shown in Table E-12.

*d. Final cell dimensions.* Use four final cells. Calculate the dimensions of the four final cells, as shown in Table E-12.

*e. Cell volume.* Calculate the primary and final cells volume, as shown in Table E-12. Allow 0.5 m (1.6 ft) for sludge storage and assume 1.2 m (4 ft) effective water depth for a total pond depth of 1.7 m. Use a 3:1 sideslopes ratio to determine the treatment volume (approximation of a frustrum).

*f. Hydraulic detention time.* Estimate the hydraulic detention time in the effective, or above-sludge, level (zone) in the primary and final cells, as shown in Table E-12.

*g. Hydraulic loading.* Check the hydraulic loading to ensure that a minimum of 75 percent total nitrogen reduction is achieved to comply with the effluent quality requirements, as shown in Table E-12.

**Table E-12**
**Constructed Wetlands Aerobic Non-Aerated Hyacinth System**

a. $BOD_5$ Loading

$BOD_5$ Loading $(kg/d)$ = Influent $BOD_5$ $(mg/L)$ × Flow $(L/d)$ × $10^{-6}$ $(kg/mg)$

$BOD_5$ Loading $(kg/d)$ = 250 $(mg/L)$ × 284 × $10^3$ $(L/d)$ × $10^{-6}$ $(kg/mg)$ = 71 $kg/d$ (157 $lb/d$)

b. Basin Surface Area

$$\text{Total Area Required } (ha) = \frac{BOD_5 \text{ Loading } (kg/d)}{\text{Moderate Loading Rate } (kg/ha\text{-}d)}$$

$$\text{Total Area Required} = \frac{71 \ (kg/d)}{50 \ (kg/ha\text{-}d)} = 1.42 \ ha \ (3.5 \ acre)$$

$$\text{Area Primary Cells } (ha) = \frac{BOD_5 \text{ Loading } (kg/d)}{\text{Primary Loading Rate } (kg/ha\text{-}d)}$$

$$\text{Area Primary Cells} = \frac{71 \ (kg/d)}{100 \ (kg/ha\text{-}d)} = 0.71 \ ha \ (1.75 \ acre)$$

c. Primary Cells Dimensions

$$\text{Cell Area } (ha) = \frac{\text{Area Primary Cells } (ha)}{2} = L \times W = L \times \frac{L}{3}$$

$$\text{Cell Area} = \frac{0.71 \ (ha)}{2} \times 10 \ 000 \ (m^2/ha) = \frac{L^2}{3}$$

L = 103 m (338 ft)
W = 34 m (112 ft)

d. Final Cells Dimensions

Divide the remaining required area (1.42 ha - 0.71 ha = 0.71 ha) into two sets of two basins each (four cells of 0.18 ha each) to produce a total system with two parallel sets with three basins each.

$$\text{Area Final Cells} = 0.18 \ (ha) \times 10 \ 000 \ (m^2/ha) = \frac{L^2}{3}$$

L = 73.5 m (241 ft)
W = 24.5 m (80 ft)

e. Cells Volume

$$V = [(L)(W) + (L - 2sd)(W - 2sd) + 4(L - sd)(W - sd)] \times \frac{d}{6}$$

(Continued)

**Table E-12 (Concluded)**

where:

    $V$ = basin volume (m³)
    $L$ = basin length at surface area (m)
    $W$ = basin width at surface area (m)
    $d$ = effective water depth = 1.2 m
    $s$ = slope factor (e.g., 3:1 slope, s = 3)

$V$ (primary cell) = 3631 m³ (128,228 ft³)

$V$ (final cell) = 1758 m³ (62,083 ft³)

*f. Hydraulic Detention Time*

$$Detention\ Time\ (d) = \frac{No.\ Cells \times Unit\ Volume\ (m^3)}{Flow\ (m^3/d)}$$

$$Detention\ Time\ (Primary)\ t_p = \frac{2 \times 3\ 631\ (m^3)}{284\ (m^3/d)} = 26\ days$$

$$Detention\ Time\ (Final)\ t_p = \frac{4 \times 1\ 758\ (m^3)}{284\ (m^3/d)} = 25\ days$$

Total detention time = 26 + 25 = 51 days at 20° C (67°F)

*g. Hydraulic Loading*

$$Hydraulic\ Loading = \frac{284\ m^3/d}{1.42\ ha} = 200\ m^3/ha{-}d$$

This hydraulic loading is within the recommended range and is sufficient (i.e., ≤ 935 m³/ha-d) to reduce the nitrogen loading by 70- 90 percent. It is reasonable to expect 5 mg/L of nitrogen or less in the effluent. Because the nitrogen will not be at optimum growth levels in this system, an annual harvest is suggested. An influent flow diffuser in each of the primary cells is recommended to properly distribute the untreated effluent.

## Appendix E
## Design Examples

### E-1.  Package Plant Extended Aeration

Design an extended aeration package plant (prefabricated or pre-engineered) to treat a municipal wastewater flow of 125 000 L/d (33 000 gal/d).  Solids retention typically ranges from 20 to 30 days; MLSS varies between 3 000 and 6 000 mg/L.  The Food to Microorganism ratio (F/M) typically varies between 0.05 to 0.30.  Influent $BOD_5$ and TSS will generally be about 250 mg/L.  The dissolved oxygen (DO) concentration is in the 1.5 to 2.5 mg/L range and preferably will never be below 2.0 mg/L.  Coarse bubble aerators will be used.  Detention time in the aeration tank will be one day.

**Table E-1**
**Design Assumptions**

Influent/Effluent Composition - Given

| Parameter | Influent | Effluent |
|---|---|---|
| $BOD_5$ | 250 mg/L | 20 mg/L |
| TSS | 250 mg/L | 20 mg/L |
| TKN | 40 mg/L | 5 mg/L |

*Assumptions*
| | |
|---|---|
| Minimum operating temperature: | 17°C (62°F) |
| Site elevation above sea level: | 137 m (450 ft) |
| Net sludge yield (kg MLSS/kg $BOD_5$): | 0.76 |
| DO mixed liquor concentration ($C_O$) | 2 mg/L |
| Oxygen coefficients: | |
| kg $O_2$/kg $BOD_5$ | 1.28 |
| kg $O_2$/kg $NH_3$-N | 4.60 |
| Transfer factors: | |
| α (typical for coarse bubble diffuser) | 0.85 |
| β (typical for domestic wastewater) | 0.95 |
| Sludge settling zone overflow rate: | < 10 m³/m²/d |
| Aeration tank detention time: | 1 day |
| Typical $O_2$ transfer rate for coarse bubble diffusers: | 30 kgO₂/kW-d (48 lbO₂/hp-d) |
| Solids retention: | 25 days |

*a.  Sludge production.*  Calculate the sludge production rate based on the desired $BOD_5$ removal.  Calculate the system total solids mass based on the sludge production rate and the assumed solids retention time, as shown in Table E-2.

*b.  Aeration power.*  Calculate the blower capacity based on the sludge production rate, desired TKN removal/synthesis, and the site specific conditions, as shown in Table E-2.

*c.  Unit dimensions.*  Estimate the required unit process dimensions including the chlorine contact tank based on one-day hydraulic detention time and using two sludge settling hoppers, as shown in Table E-2.

**Table E-2**
**Package Plant Extended Aeration Design Calculations**

a. Aerobic Volume

$$BOD_5 \ Removed \ (kg/d) = \frac{Flow \ (L/d)}{10^6 \ (mg/kg)} \times (BOD_{influent} - BOD_{effluent}) \ (mg/L)$$

$$BOD_5 \ Removed = \frac{125 \times 10^3}{10^6} \times (250 - 20) \cong 29 \ kg/d \ (64 \ lb/d)$$

$$Sludge \ Production \ (kg/d) = Net \ Sludge \ Yield \ (kg \ MLSS/kg \ BOD_5) \times BOD_5 \ Removed \ (kg/d)$$

$$Sludge \ Production = 0.76 \times 29 = 22 \ kg/d \ (48.5 \ lb/d)$$

$$System \ Mass \ (kg) = Sludge \ Production \ (kg/d) \times Solids \ Retention \ (d)$$

$$System \ Mass = 22 \ (kg/d) \times 25 \ (d) = 550 \ kg \ (1 \ 213 \ lb)$$

b. Aeration Power

$$NH_3 - N_{oxidized} = TKN_{influent} - Synthesis \ N - TKN_{effluent}$$

$$Synthesis \ N = 5\% \ waste \ activated \ sludge \ of \ total \ daily \ sludge \ production$$

$$Synthesis \ N = 0.05 \times 22 \ (kg/d) = 1.1 \ kg/d \ (2.4 \ lb/d)$$

$$Synthesis \ N \ (mg/L) = \frac{1.1 \ (kg/d) \times 10^6 \ (mg/kg)}{125 \times 10^3 \ (L/d)} = 8.8 \ mg/L$$

$$NH_3 - N_{oxidized} = 40 \ (mg/L) - 8.8 \ (mg/L) - 5 \ (mg/L) = 26.2 \ mg/L$$

$$NH_3-N \ (kg/d) = 26.2 \ (mg/L) \times 10^{-6} \ (kg/mg) \times 125 \times 10^3 \ (L/d) = 3.28 \ kg/d \ (7.2 \ lb/d)$$

$$AOR = 1.28 \ (kgO_2/kgBOD_5) \times Synthesis \ N \ (kgBOD_5/d) + \\ 4.6 \ (kgO_2/kg \ NH_3-N) \times NH_3-N_{oxidized} \ (kg/d)$$

$$AOR = 1.28 \ (kgO_2/kgBOD_5) \times 1.1 \ (kgBOD_5/d) + 4.6 \ (kgO_2/kg \ NH_3-N) \times 3.28 \ (kgNH_3-N/d)$$

$$AOR = 16.5 \ kgO_2/d \ (36.4 \ lbO_2/d)$$

where:

AOR = Actual Oxygen Requirements (kg $O_2$/d)

$$SAOR = AOR \times \frac{C_S \ (mg/L) \times \Theta^{(20-T)}}{\alpha \times (\beta \times C_{SW} - C_O)}$$

where:

SAOR = Standard Actual Oxygen Requirements (kg $O_2$/d)
$\Theta$ (temperature correction factor) = 1.024

(Sheet 1 of 3)

**Table E-2 (Continued)**

$C_S$ ($O_2$ saturation concentration at standard temperature and pressure) = 9.02 mg/L
$C_{SW}$ = correction factor for elevation ( i.e., 450 ft) = 9.02 - 0.0003 × elevation
$C_{SW}$ = 9.02 - 0.0003 × 450 = 8.88 mg/L (<u>NOTE:</u> 0.0003 may be used as rule-of-thumb describing a 0.0003 mg/L rise/drop in DO saturation concentration per every foot of elevation increase/decrease.)
$C_O$ = 2 mg/L
$\alpha$ = 0.85; $\beta$ = 0.95; T = 17°C (62°F)

$$SAOR = 16.5 (kgO_2/d) \times \frac{9.02\ (mg/L) \times 1.024^{(20-17)}}{0.85 \times [0.95 \times 8.88\ (mg/L) - 2.0\ (mg/L)]} = 29.2\ kgO_2/d\ (64.4\ lbO_2/d)$$

$$Motor\ Requirements\ (kW) = \frac{SAOR\ (kgO_2/d)}{O_2\ Transfer\ Rate\ (kgO_2/kW\text{-}d)}$$

$$Motor\ Requirements\ (kW) = \frac{29.2\ (kgO_2/d)}{30\ (kgO_2/kW\text{-}d)} = 1.0\ kW\ (1.3\ hp)$$

Since blowers typically have an efficiency of 50% or less, select 2 aerators with 3.73 kW (5 hp) motors.

c. Unit Dimensions

1. Aeration Tank Volume = 125 $m^3$ at one day hydraulic detention

   Assume tank dimensions based on values typical of extended aeration systems:
   Operating depth = 3.048 m (10 ft)
   Width = 2.895 m (9.5 ft)

$$Tank\ Length = \frac{Volume}{Width \times Depth} = \frac{125}{2.895 \times 3.048} = 14.2\ m\ (47\ ft)$$

2. Sludge settling zone overflow rate. To calculate the sludge settling zone overflow, assume two sludge settling hoppers each with a top dimension of 2.895 m (9.5 ft) and a bottom dimension of 0.304 m (1 ft).

   Surface area of sludge settling zone = 2 × 2.895 × 2.895 = 16.76 $m^2$ (180 $ft^2$)

$$Overflow\ Rate = \frac{125\ (m^3/d)}{16.76\ (m^2)} = 7.46\ m^3/m^2\text{-}d\ (24.5\ ft^3/ft^2\text{-}d)$$

   Assume settling height above hoppers = 0.3 m (1 ft)

   Depth of hopper = 3.048 m - 0.3 m = 2.75 m (9 ft)

$$Hopper\ Volume = \frac{1}{3}\ (A_1 + A_2 + \sqrt{A_1 \times A_2}\ ) \times depth$$

$$Hopper\ Volume = \frac{1}{3}\ (2.895^2 + 0.304^2 + \sqrt{2.895^2 \times 0.304^2}\ ) \times 2.75 = 8.56\ m^3\ (302\ ft^3)$$

   Total Hopper Volume = 8.56 $m^3$ x 2 = 17.1 $m^3$ (604 $ft^3$)

   Sludge holding tank shall be located at head of tank and shall equal volume of sludge hoppers.

   Width = 2.895 m (9.5 ft)
   Depth = 3.048 m (10 ft)

$$Holding\ Tank\ Length\ (m) = \frac{Total\ Hopper\ Volume\ (m^3)}{Width\ (m) \times Depth\ (m)}$$

$$Holding\ Tank\ Length = \frac{17.1\ m^3}{2.895\ m \times 3.048\ m} = 1.94\ m\ (6.4\ ft)$$

(Sheet 2 of 3)

---

**Table E-2 (Concluded)**

3. If chlorination is used as a disinfectant, the chlorine contact tank shall have a detention time of 75 min; therefore the tank shall have a capacity of 6.5 m³ and the tank dimensions will be:

Width = 2.895 m (9.5 ft)
Length = 3.048 m (10 ft)

$$\textit{Chlorine Contact Tank Depth (m)} = \frac{\textit{Volume } (m^3)}{\textit{Width } (m) \times \textit{Depth } (m)} = \frac{6.5 \ m^3}{2.895 \ m \times 3.048 \ m}$$

Chlorine Contact Tank Depth = 0.74 m (2.4 ft)

(Sheet 3 of 3)

---

*d.   Equipment specifications.* Figure E-1 presents a plan view and a side view of the pre-engineered package plant extended aeration with the following specifications:

(1)   The unit package plant will require no pre-treatment other than wastewater pumping from an influent manhole lift station.

(2)   The influent pipe shall have a minimum of a 150 mm (6 in.) diameter from the influent manhole and will discharge directly to a combination comminutor/bar screen located ahead of (and on top of) the aeration tank.

(3)   Two 3.73 kW (5 hp) blower assemblies shall provide air at 31 kPa (4.5 psi) to ensure a 2.0 mg/L DO level in the aeration tank at all times.

(4)   A minimum of 44 diffusers will be required to distribute aeration at the aeration tank floor level. At least six (6) diffusers will be provided in the sludge holding tank and one (1) in the chlorine contact tank.

(5)   A totalizing flow meter will be provided to record the daily flow patterns and total.

(6)   A minimum of eight spray nozzles will be required on the top of the aeration tank on the side opposite to the aeration diffuser drops.

(7)   Each sludge hopper will be equipped with an air lift pump with openings 150 mm (6 in.) above the hopper bottoms.

(8)   The air lift pumps will discharge to a combination 75 mm (3 in.) sludge return and sludge waste line to the head of the tank.

(9)   Blower units shall be controlled by a blower panel located above the aeration tank.

(10)   Scum skimmers will be provided at a scum baffle ahead of the tank discharge (by V-notch weir) to the chlorine contact tank.

(11)   Should ultraviolet disinfection be chosen in lieu of chlorination of tank effluent, an in-pipe rather than open channel effluent flow may be specified.

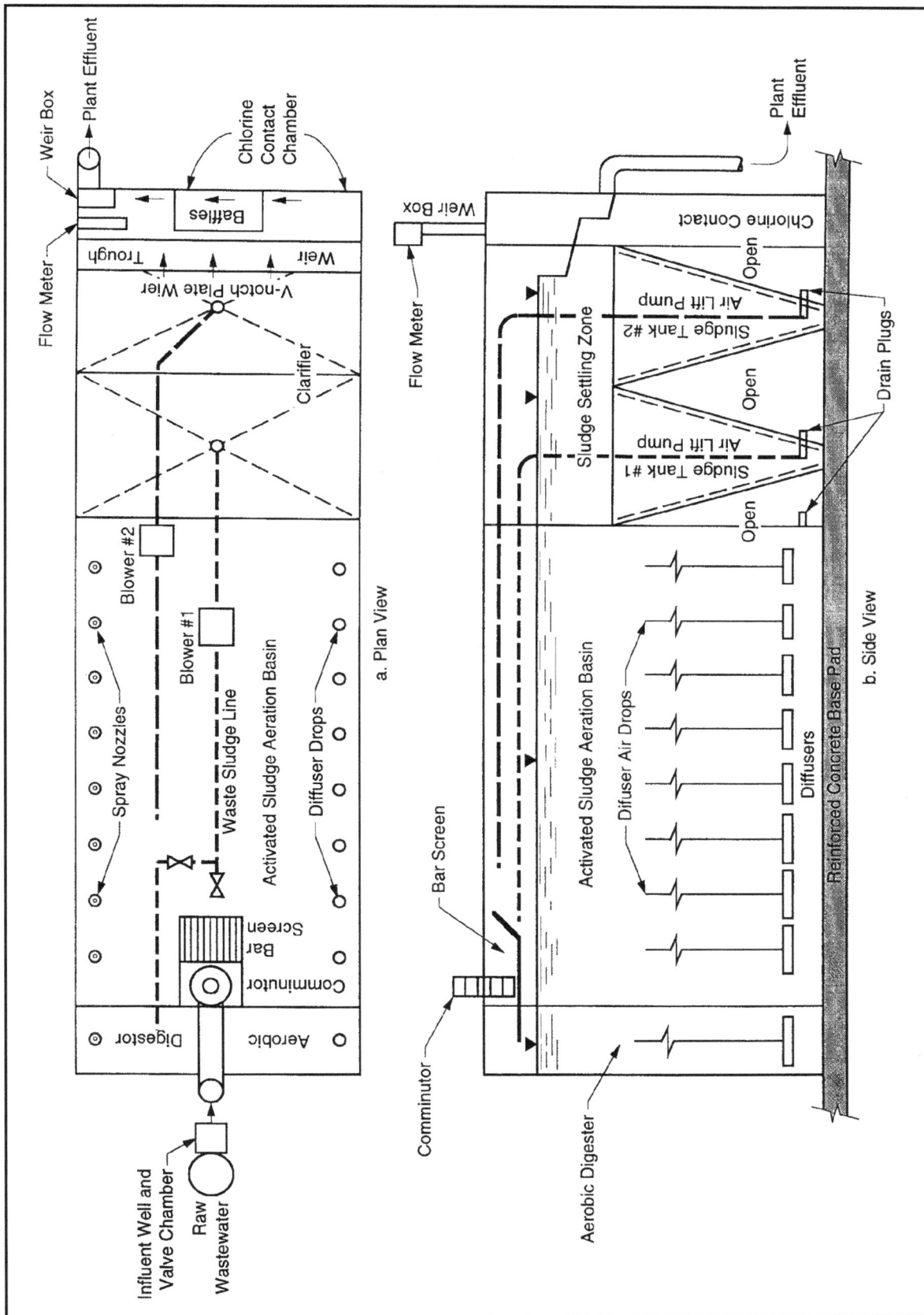

Figure E-1. Pre-engineered package plant extended aeration

## E-2. Oxidation Ditch (Continuous-Loop Reactor) Carrousel—Wraparound

Design a carrousel (circular or wraparound) oxidation ditch to treat municipal wastewater at an average influent flow rate of 378 500 L/d (100,000 gal/d). The new system will use mechanical aerators and have the design parameters shown in Table E-3.

**Table E-3**
**Design Parameters and Assumptions**

*Influent/Effluent Composition*

| Parameter | Influent | Effluent |
|---|---|---|
| $BOD_5$ | 250 mg/L | 5 mg/L |
| TSS | 300 mg/L | 10 mg/L |
| TKN | 30 mg/L | 5 mg/L |
| $NH_3$-N | -- | 0.5 mg/L |

*Assumptions*

| | |
|---|---|
| Minimum wastewater temperature: | 16°C (61°F) |
| Process solids retention time: | 20 days |
| MLSS concentration: | 4000 mg/L |
| Net yield (kg MLSS/ kg $BOD_5$): | 0.76 |
| Oxygen coefficients: | |
| kg $O_2$/kg $BOD_5$ | 1.28 |
| kg $O_2$/kg $NH_3$-N | 4.60 |
| Transfer factors: | |
| $\alpha$ (typical for mechanical aerator) | 0.90 |
| $\beta$ (typical for domestic wastewater) | 0.95 |
| Typical $O_2$ transfer rate for mechanical aerator: | |
| | 37 kg$O_2$/kW-d |
| Site elevation (sea level + tank height): | (60 lb$O_2$/hp-d) |
| Clarifier overflow rate: | 9.1 m (30 ft) |
| Side water depths, | 16.3 m³/m²/d |
| clarifier and reactor channels: | 3.04 m (10 ft) |

   *a.  Carrousel volume.* Calculate the sludge production rate based on the desired $BOD_5$ removal. Calculate the system total solids mass based on the sludge production rate and the assumed solids retention time. Calculate the carrousel volume from the calculated system total solids mass and the assumed MLSS concentration, as shown in Table E-4.

   *b.  Aeration power.* Calculate the blower capacity based on the desired TKN removal/synthesis and the site specific conditions, as shown in Table E-4.

   *c.  Clarifier diameter.* Estimate the required wraparound clarifier diameter based on the assumed clarfier overflow rate and the side water depths, as shown in Table E-4.

   *d.  Carrousel specifications.* The carrousel shown in Figure E-2 has the following specifications:

| | |
|---|---|
| Clarifier diameter: | 6.4 m (21 ft) |
| Inner channel: | 4 m (13 ft) |
| Outer channel: | 4 m (13 ft) |
| Entire tank diameter: | 14 m (46 ft) |
| Walls and miscellaneous equipment thickness: | 2 m (6.5 ft) |
| Constructed carrousel diameter: | 14 m + 2 m = 16 m (52.5 ft) |

**Table E-4**
**Oxidation Ditch Design Calculations**

a. Carrousel Volume

$$BOD_5 \ Removed \ (kg/d) = \frac{Flow \ (L/d)}{10^6 \ (mg/kg)} \times (BOD_{influent} - BOD_{effluent}) \ (mg/L)$$

$$BOD_5 \ Removed = \frac{378.5 \times 10^3}{10^6} \times (250 - 5) = 92.7 \ kg/d \ (204 \ lb/d)$$

$$Sludge \ Production \ (kg/d) = Net \ Yield \ (kg \ MLSS/kg \ BOD_5) \times BOD_5 \ Removed \ (kg/d)$$

$$Sludge \ Production = 0.76 \times 92.7 = 70.5 \ kg/d \ (156 \ lb/d)$$

$$System \ Mass = 70.5 \ (kg/d) \times 20 \ (d) = 1\ 410 \ kg \ (3\ 107 \ lb)$$

$$Carrousel \ Volume \ (m^3) = \frac{System \ Mass \ (kg) \times 10^3}{MLSS \ Concentration \ (mg/L)}$$

$$Carrousel \ Volume = \frac{1.41 \times 10^3 \times 10^3}{4 \times 10^3} = 353 \ m^3 \ (12,466 \ ft^3)$$

b. Aeration Power

*Synthesis N =5% wasted activated sludge of total daily sludge production*

*Synthesis N = 0.05 × 70.5 (kg/d) = 3.52 kg/d (7.75 lb/d)*

*Synthethis N = 9.3 mg/L (for a daily flow of 378 500 L/d)*

$$NH_3 - N_{oxidized} = TKN \ (mg/L) - Synthesis \ N \ (mg/L) - N\text{-}NH_{3(effluent)} \ (mg/L)$$

$$NH_3 - N_{oxidized} = 30 - 9.3 - 0.5 = 20.2 \ mg/L$$

$$NH_3 - N_{oxidized} = 7.6 \ kg/d \ (16.9 \ lb/d) \text{ - based on a daily flow of 378 500 L/d}$$

$$AOR = kg \ O_2/kg \ BOD_5 \times Synthesis \ N \ (kg \ BOD_5/d) + kg \ O_2/kg \ NH_3\text{-}N \times NH_3\text{-}N_{oxidized} \ (kg/d)$$

$$AOR = 1.28 \times 3.52 + 4.6 \times 7.6 = 4.5 + 35.0 = 39.5 \ kg \ (87 \ lb)$$

where:
AOR = Actual Oxygen Requirements (kg O₂/d)

$$SAOR = AOR \times \frac{C_S \ (mg/L) \times \Theta^{(20-T)}}{\alpha \times (\beta \times C_{SW} - C_O)}$$

$$Clarifier \ Area \ (m^2) = \frac{Design \ Flow \ (m^3/d)}{Overflow \ Rate \ (m^3/m^2/d)}$$

where:

SAOR = Standard Actual Oxygen Requirements (kgO₂/d)
Θ (temperature correction factor) = 1.024
$C_S$ (DO saturation concentration at standard temperature and pressure conditions) = 9.02 mg/L
$C_{SW}$ = Correction factor for elevation (i.e., 30 ft) = 9.02 - 0.0003 × elevation
$C_{SW}$ = 9.02 - 0.0003 × 30 = 9.011 mg/L (NOTE: 0.0003 may be used as rule-of-thumb describing a 0.0003 mg/L rise/drop in DO saturation concentration per every foot of elevation increase/decrease.)

(Sheet 1 of 3)

**Table E-4 (Continued)**

$C_O$ = 2.0 mg/L
$\alpha$ = 0.90;  $\beta$ = 0.95;  T = 16°C (61°F)

$$SAOR = 39.5 \ (kgO_2/d) \times \frac{9.02 \ (mg/L) \times 1.024^{(20-16)}}{0.90 \times [0.95 \times 9.01 \ (mg/L) - 2.0 \ (mg/L)]}$$

$$= 66.4 \ kg \ O_2/d \ (146 \ lb \ O_2/d)$$

$$Aerator \ Power \ Requirements \ (kW) = \frac{SAOR \ (kgO_2/d)}{O_2 \ Transfer \ Rate \ (kgO_2/kW\text{-}d)}$$

$$Aerator \ Power \ Requirements = \frac{66.4 \ (kgO_2/d)}{37 \ (kgO_2/kW\text{-}d)} = 1.80 \ kW \ (2.4 \ hp)$$

Since blowers typically have an efficiency of 50% or less, select two aerators with 3.73 kW (5 hp) motors or two 2-speed aerators with 7.5/5 hp motors per basin normally operated on low speed.

c. Clarifier Diameter

Design flow = 387 500 L/d

Overflow rate = 16.3 m³/m²/d

$$Clarifier \ Area \ (m^2) = \frac{Design \ Flow \ (m^3/d)}{Overflow \ Rate \ (m^3/m^2/d)}$$

$$Clarifier \ Area = \frac{378.5 \ (m^3/d)}{16.3 \ (m^3/m^2/d)} = 23.2 \ m^2 \ (250 \ ft^2)$$

$$Clarifier \ Surface \ Area = \frac{\pi D^2}{4} = 23.2 \ m^2$$

Solve for diameter (D):

$$D = \sqrt{23.2/0.785} \cong 6.0 \ m \ (20 \ ft)$$

A new clarifier surface area is then recalculated:

$$New \ Clarifier \ Area \ (m^2) = \frac{\pi \times D^2}{4}$$

$$New \ Clarifier \ Area = \frac{\pi \times [6 \ (m)]^2}{4} = 28.3 \ m^2 \ (305 \ ft^2)$$

Detention time (hr) = Volume (m³) × 24 (hr/d)/Flow (m³/d)

Volume = Area (m²) × Depth (m) = 28.3 (m²) × 3.04 (m) = 86 m³ (3 038 ft³)

(Sheet 2 of 3)

Table E-4  (Concluded)

$$Detention\ time = \frac{86\ (m^3) \times 24\ (hr/day)}{378.5\ (m^3/d)} = 5.5\ hr$$

Total clarifier diameter = water diameter + 2 × wall thickness

Total clarifier diameter = 6 m + 2 × 0.2 m = 6.4 m (21 ft)

For a 3.04 m side water depth (SWD) for the channels

$$Volume = 378.5\ m^3\ of\ wraparound\ channels$$

$$A = \frac{\pi D_1^2}{4} - \frac{\pi D_2^2}{4} = 378.5\ m^3/SWD$$

$$0.785\ D_1^2 - 0.785\ D_2^2 = 378.5/3.04$$

$$0.785\ D_1^2 - 0.785 \times 6.4^2 = 124.5$$

D₁ = 14 m (46 ft)

(Sheet 3 of 3)

## E-3.  Stabilization Pond

Design a facultative stabilization pond with primary treatment (clarifier and anaerobic digester of Imhoff tank design) to be followed by secondary clarification to treat a domestic wastewater flow of 378 500 L/d (100,000 gal/d).  Influent $BOD_5$ will be 250 mg/L.  Assume a primary clarifier removes 33 percent of the influent $BOD_5$ ($BOD_5 = 0.68\ BOD_u$), and influent wastewater $[SO_4^{2-}]$ is $\leq 500$ mg/L.  Four rectangular ponds in parallel are to be constructed.  The controlling winter temperature of each pond will be 4.5 °C (40 °F).  Length to width ratio of each pond will be 3:1, as is typical for such facilities.

Find, as shown in Table E-5:

Total area of ponds.

Applied $BOD_5$ loading.

Dimensions of the ponds.

16 m

3.0 m

6.4 m

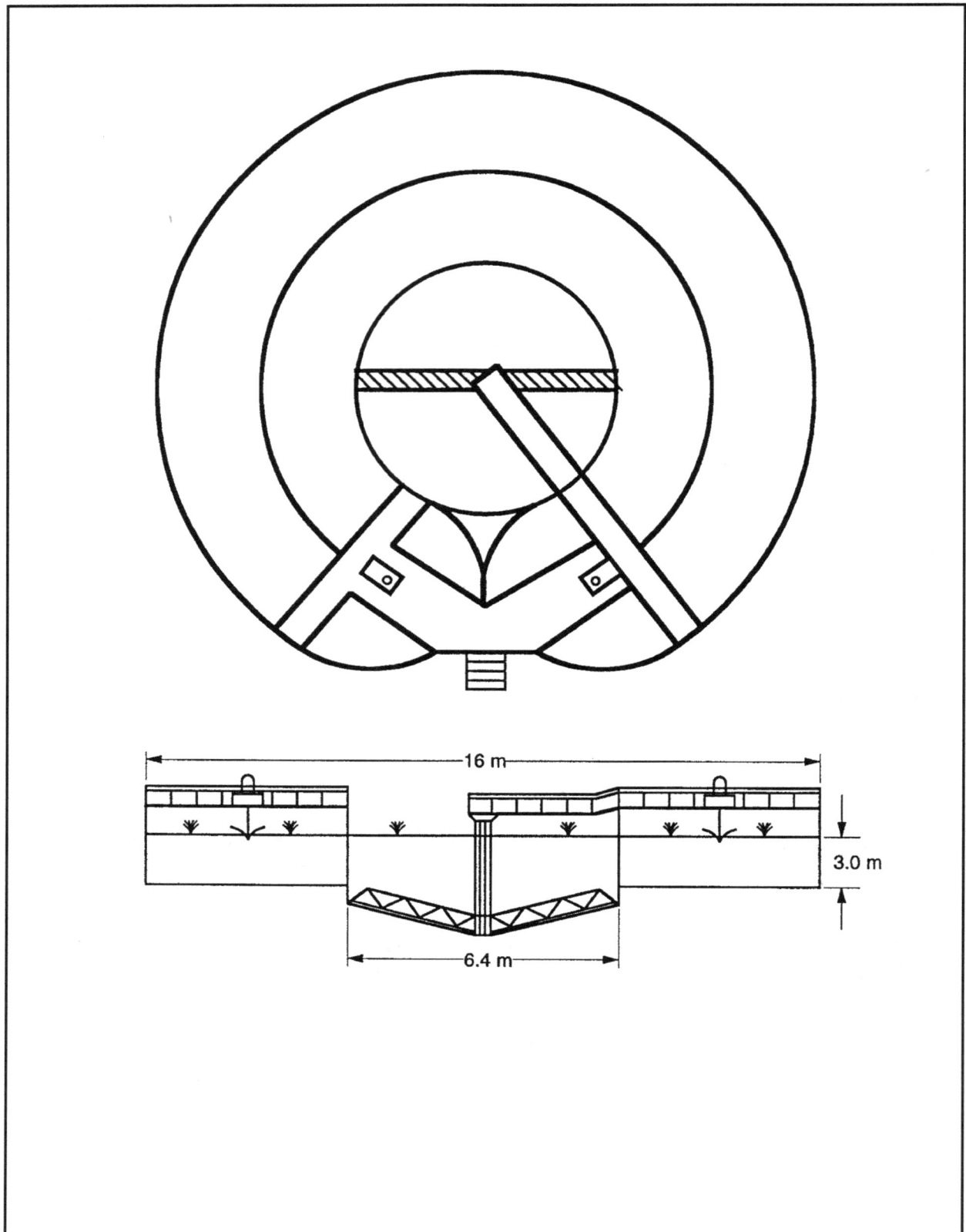

Figure E-2. Oxidation ditch carrousel wraparound (closed-loop reactor)

E-10

**Table E-5**
**Stabilization Pond Design Calculations**

$BOD_5$ (primary clarifier effluent) = 250 (mg/L) × (1-0.33) = 167.5 mg/L

$S_i = BOD_u = 167.5/0.68 = 246.3$ mg/L

$t = \dfrac{V}{Q} = 0.035 \times 246.3 \times [1.085^{(35-4.5)}] \times 1 \times 1 = 104$ days

$V$ (4 basins) $= Q$ ($m^3/d$) $\times t$ ($d$) $= 378.5$ ($m^3/d$) $\times 104$ ($d$) $= 39\ 364\ m^3$ (31.9 acre-ft)

$A$ (4 basins) $= \dfrac{V}{d}$

where:

A = total ponds area (hectares)
d = facultative pond effective depth = 1.8 m (6 ft); to include 0.3 m (1 ft) for sludge storage

$A$ (4 basins) $= \dfrac{39\ 364\ (m^3)}{1.8\ (m)} \times \dfrac{1\ ha}{10\ 000\ (m^2)} = 2.19\ ha$ (5.41 ac)

$BOD_5$ Load $= 387.5 \times 10^3$ (L/d) $\times 167.5$ (mg/L) $\times 10^{-6}$ (kg/mg) $= 64.9$ kg/d (142.9 lb/d)

Applied Load $= 64.9$ (kg/d) $\times \dfrac{1}{2.19\ ha}$

Applied Load = 29.6 kg $BOD_5$/ha-d (26.5 lb $BOD_5$/acre-d)

Each Pond Area $= \dfrac{\text{Total Pond Area (ha)} \times 10\ 000\ (m^2/ha)}{4} = 5\ 475\ m^2$ (17 958 $ft^2$)

Pond Area = Length × Width

Length = 3 × width (W)

Pond Area = (3W) × W = 5 475 $m^2$

Width (W) = 42.7 m (140 ft)

Length = 128 m (420 ft)

Use the Gloyna equation ("Facultative Waste Stabilization Pond Design in Ponds as a Wastewater Treatment Alternative," by E. F. Gloyna, J. F. Malina, Jr. and E. M. Davis):

$$t = \frac{V}{Q} = CS_i\ [\theta^{(35-T)}]ff'$$

where

$V$ = pond volume, m³.

$C$ = conversion coefficient = 0.035 (a constant metric conversion).

$Q$ = flow (378.5 m³/d).

$S_i$ = ultimate influent $BOD_5$ mg/L.

$f$ = sulfide or immediate chemical oxygen demand = 1 (for[ $SO_4^{2-}$] concentrations $\leq 500$ mg/L).

$f$ = algae toxicity factor = 1.

$\theta$ = temperature coefficient (the value of $\theta$ ranges from 1.036 to 1.085, and 1.085 is recommended as it is conservative).

$T$ = average water temperature for the pond during winter months, °C.

$t$ = hydraulic detention time (days).

## E-4. Zero Discharge or Water Recycle/Reuse For Toilet Flush Water in Rest Areas (Closed-Loop Reuse)

*a. Background.*

(1) A combination of an extended aeration-activated sludge wastewater treatment system followed by mixed-media pressure filtration has been successful in treating liquid waste from a comfort facility with eight water closets and two urinals, plus lavatories. The design wastewater flow is 37 800 L/d (10,000 gal/d).

(2) The closed-loop reuse principle is generally applicable where liquid discharges from a recreational area are not permitted or desired. After the system is initially filled and operational, a fraction of the treated wastewater (about 6 percent) is fed to the terminal holding pond or lagoon to evaporate to account for the makeup water used for lavoratories and drinking fountains. The makeup water is estimated to represent about 6 percent of total water use. The sludge from the waste solids holding basin is periodically removed by tank truck. The design parameters for the original extended aeration treatment system are presented in Table E-6. Figure E-3 presents a schematic flow diagram of the wastewater recycle-reuse system.

(3) It is to be expected that 90 to 95 percent of water used in a comfort station facility is for the water closets or toilet flushing functions. Generally, 10 to 20 cycles are required for the system to reach equilibrium with an input of 5 to 10 percent of potable water for the lavatories or drinking fountains. The wastewater from lavatories and drinking fountains is considered "new" water and is a factor in the control of the amount of wastewater that must be fed to the final holding pond to evaporate.

(4) Operating records reveal no objectionable odors from the water closets or lavatories, no objectionable colors from blue (or other food dyes) introduced to give a sanitized look to the flushing waters, no foaming in the sanitary facilities, and no building of total suspended solids. The 90 to 95 percent of reused water in the water closets and urinals has an acceptable quality following the extended aeration process and multimedia filtration.

(5) Use surveys to indicate that toilet flush water use is about 12.7 L (3.4 gal) per flush and 15.0 L (4.0 gal) per toilet user. Potable water use (lavatories and drinking fountains) is approximately 0.8 L (0.2 gal) per toilet user. Average resident time in the toilet facility is expected to be 3 min.

*b. Recycled wastewater.*

(1) The desired treatment characteristics of the recycled wastewater are shown in Table E-7.

**Table E-6**
**Original Design Parameters of the Existing Extended-Aeration System**

| Parameter | Value |
|---|---|
| Design flow: | 37 800 L/d (10,000 gal/d) |
| Aeration Tank: | |
| Detention time | 24 hr |
| Volume | 38 m³ (1342 ft³) |
| Oxygen transfer rate | 756 g/hr (1.7 lb/hr) |
| Max. Return Solids flow: | 1.83 L/s (0.48 gal/s) |
| Waste solids holding basin: | 15.1 m³ (533 ft³) |
| Comminutor: | 8.8 L/s (2.3 gal/s) |
| Settling basin: | |
| Detention time | 4 hr |
| Volume | 6.3 m³ (222 ft³) |
| Holding pond: | |
| Volume | 567 m³ (20,000 ft³) |
| Surface area | 497 m² (5350 ft²) |

The comminutor shreds to 6 mm (0.2 in). An overflow bypass around the comminutor has a manually cleaned medium bar screen. All pumps are pneumatic.

**Table E-7**
**Desired Characteristics of Recycled Wastewater**

| Parameter | Concentration Range |
|---|---|
| MLSS | 3000 to 5000 mg/L |
| Settleability | 200 to 850 mL |
| (as determined by the MLSS volume in 1 L graduated cylinder after 1 hr) | |
| Alkalinity | |
| pH | |
| TSS | 50 to 500 mg/L |
| | 5.5 to 8.3 |
| | < 15 mg/L |

(2) To achieve the best operation, the recycled wastewater must be chemically stable and the total suspended solids and total volatile solids must remain relatively constant. The most desirable range for MLSS would probably be 3500 to 4000 mg/L with an accompanying settleability of 400 to 600 mL.

*c. Unit processes for closed-loop reuse.*

(1) The unit processes shown in Table E-8 have been added for the closed-loop reuse to meet the desired characteristics identified in Table E-7.

(2) The multimedia rapid filtration pressurized vessel has a design filtration rate of 80 to 160 L/min/m² and a backwash design flow rate of 285 to 610 L/min/m². The filter appears to operate best at a filtration rate of 94 L/min/m² (2.3 gal/min/ft²) and at a backwash cleaning rate of 345 L/min/m² (8.5 gal/min/ft²). Total suspended solids in the recycled wastewater must be less than 15 mg/L for reuse in the toilet facility.

## E-5. Sequencing Batch Reactor (SBR)

*a. General.*

(1) The design of a sequencing batch reactor (SBR) involves the same factors commonly used for the flow-through activated sludge system. The aspects of a municipally treated waste which require

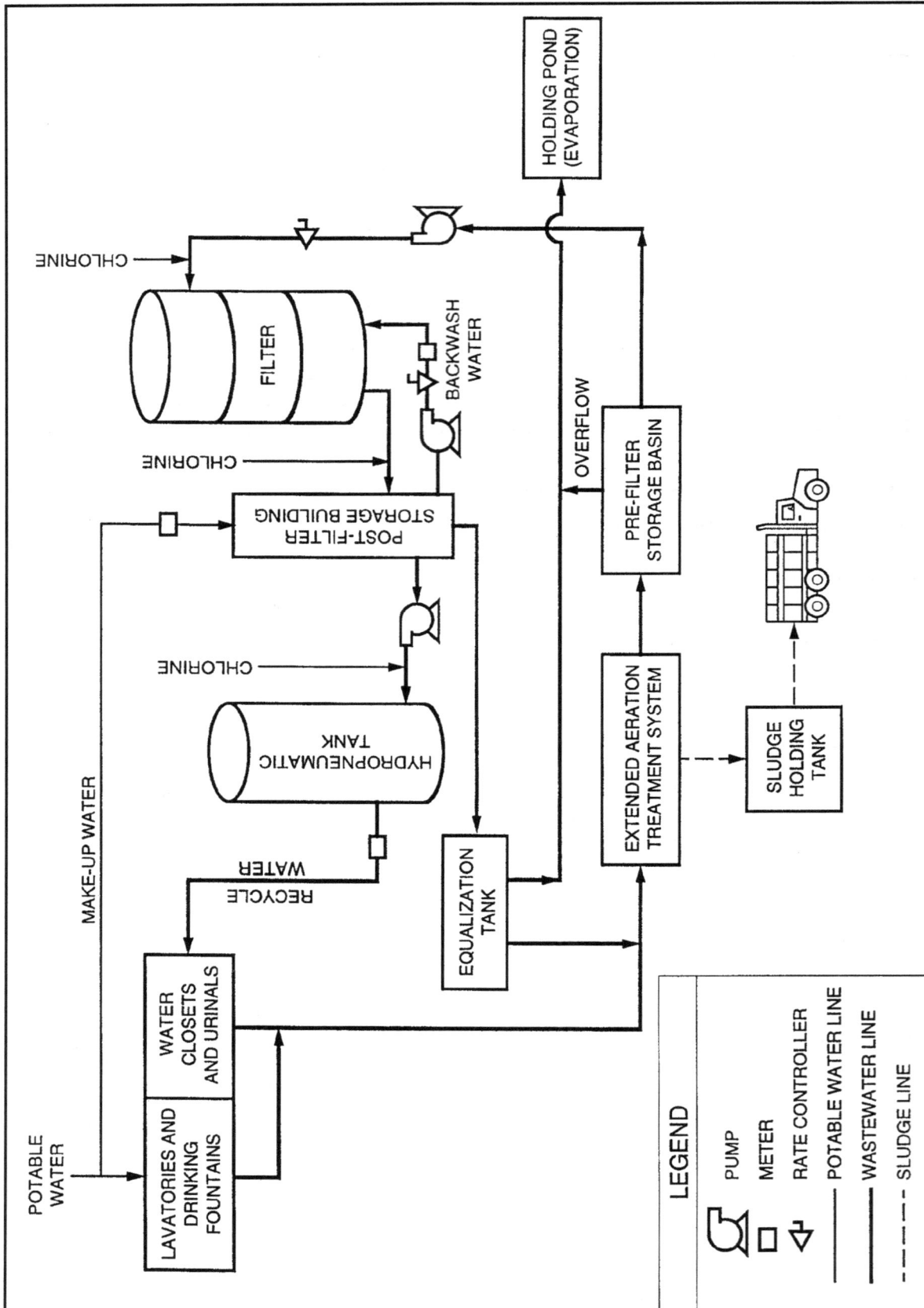

Figure E-3. Flow diagram for wastewater recycle-reuse system

**Table E-8**
**Unit Processes for the Closed Loop System**

| Unit Process | Design Parameter |
|---|---|
| Pressure Filter | |
|   Diameter | 1.8 m (6 ft) |
|   Media | Granular nonhydrous aluminum silicate |
| | (Effective size = 0.57, |
| | Uniform coefficient = 1.66) |
| | 4.1 L/s (65 gal/min) |
|   Max. pump rate | 1.6 L/s/m$^2$ (0.04 gal/s/ft$^2$) |
|   Filtration rate | 2.65 L/s/m$^2$ (0.06 gal/s/ft$^2$) |
|   Surface wash rate | 5.77 L/s/m$^2$ (0.13 gal/s/ft$^2$) |
|   Backwash rate | |
| Pre-Filter Storage Tank | 75.6 m$^3$ (2670 ft$^3$) |
| Post-Filter Storage Tank | 75.6 m$^3$ (2670 ft$^3$) |
| Equalization Tank | 18.9 m$^3$ (668 ft$^3$) |
| Hydropneumatic Tank | |
|   Total volume | 18.9 m$^3$ (668 ft$^3$) |
|   Operating volume | 5.3 m$^3$ (187 ft$^3$) |

dentrification as well as nitrification plus biological phosphorous removal need additional design considerations. Pretreatment of the wastewater before influent in the SBR reactor system is also required.

(2) The following example should be considered an outline to identify reactor volume elements, a diffused aeration system, the basis for signing effluent decanter units, and waste sludge systems for a system receiving 378 500 L/d (100,000 gal/d) of wastewater.

(3) Food-to-mass (F/M) ratio typically ranges from 0.05 to 0.30 with domestic waste F/M ratios typically ranging from 0.10 to 0.15. At the end of the decant phase, the MLSS concentration may vary between 2000 and 5000 mg/L. A typical value for a municipal waste would be 3500 mg/L. The MLSS concentration changes continuously throughout an SBR operating cycle from a maximum at the beginning of a fill phase to a minimum at the end of the react phase.

*b. Reactor volume.* Calculate the reactor volume based on the desired BOD$_5$ removal, the F/M ratio, and the MLSS. The F/M ratio and the MLSS at the low water level determine the reactor volume at the low water level, as shown in Table E-10.

*c. Decant volume.* Calculate the decant volume as the difference between the reactor volume and the low water volume, as shown in Table E-10. Each operating cycle is normally composed of *mixed fill*, *react fill*, *settle*, *decant*, *sludge waste*, and *idle*. The number of cycles dictates the number of decants per day or the volume of liquid to be decanted for each cycle. The volume per decant per cycle must be selected based on the maximum sustained daily flow.

*d. Detention time.* Calculate the maximum detention time based on the reactor volume. Calculate the minimum detention time based on the decant volume, as shown in Table E-10.

*e. SBR dimensions.* Estimate the required unit process dimensions, as shown in Table E-10. The basin length can be estimated based on a recommended minimum depth. The minimum depth after decant is determined as the depth of a clarifier in a flow-through system, i.e., quiescent settling and a large settling area. A minimum depth of 2.75 m (9 ft) is typically recommended by designers.

**Table E-9**
**Design Assumptions**

*Influent/Effluent Composition*

| Parameter | Influent | Effluent |
|---|---|---|
| BOD5 | 250 mg/L | 25 mg/L |
| TSS | 250 mg/L | 30 mg/L |
| NH3-N | 25 mg/L | 1 mg/L |
| Total Phosphorous | 10 mg/L | 2 mg/L |
| TKN | 40 mg/L | 5 mg/L |

*Assumptions*

| | |
|---|---|
| F/M Ratio (kgBOD$_5$applied/kgMLSS-d): | 0.13 |
| MLSS: | 3500 mg/L |
| Minimum clarifier depth: | 2.75 m (9 ft) |
| Net sludge yield (kg MLSS/kg BOD$_5$): | |
| Wastewater temperature: | 0.76 |
| Minimum solids retention time: | 20°C(67°F) |
| Reactor volume decanted each day: | 8 d |
| Net elevation above sea level: | 60% |
| DO mixed liquor concentration (C$_o$): | 304 m (1000 ft) |
| Oxygen coefficients: | 2 mg/L |
| kg O$_2$/kg BOD$_5$ | |
| kg O$_2$/kg NH$_3$-N | 1.28 |
| Transfer factors: | 4.60 |
| α (typical for coarse bubble diffuser) | |
| β (tyical for domestic wastewater) | 0.85 |
| Typical O2 transfer rate for coarse bubble diffusers: | |
| | 0.95 |
| Number of cycles per day: | |
| Include two square basins for operational flexibility | 1.25 kgO$_2$/kW-hr (2 lbO$_2$/hp-hr) 4 |

*f.* *Aeration power.* Calculate the blower capacity based on the sludge production rate and the total oxygen demand, as shown in Table E-10. In sizing the aeration equipment, it must be noted that the equipment operates only a portion of the SBR operating cycle (part of the fill plus react phases). Therefore, the total daily oxygen requirements must be met in a shorter time period than in a conventional activated sludge flow-through system. The total daily oxygen requirements are estimated by adding the carbonaceous oxygen demand (oxygen required for BOD$_5$ oxidation) to the nitrogeneous oxygen demand (oxygen required for TKN oxidation).

*g.* *Sludge and decanter flows.* Calculate the sludge and decanter flow rates at design conditions, as shown in Table E-10.

*h.* *Equipment specifications.* Figure E-4 presents a plan view of the SBR system with the following specifications:

(1) At least two basins are provided in an SBR design to provide operational flexibility and improved effluent quality. SBR unit dimensions:

maximum volume = 470 m³ (16,598 ft³).
high water level = 5.80 m (19 ft).
maximum decant height = 3.05 m (10 ft).
low water level = 2.75 m (9 ft).
hydraulic detention time at low water level = 17.9 hrs.

**Table E-10**
**SBR Design Calculations**

*a. Reactor Volume*

$$BOD_5 \ Removed \ (kg/d) = [(BOD_{influent} - BOD_{effluent}) \ (mg/L)] \times Flow \ (L/d) \times 10^{-6} \ (kg/mg)$$

$$BOD_5 \ Removed = (250 - 25) \times 378.5 \times 10^{-3} = 85.2 \ kg/d \ (187 \ lb/d)$$

$$Required \ Aerobic \ Mass \ (kg) = \frac{BOD_5 \ Removed \ (kg/d)}{F/M \ Ratio \ (kgBOD_5/kbMLSS-d))}$$

$$Required \ Aerobic \ Mass = \frac{85.2 \ (kgBOD_5/d)}{0.13 \ (kgBOD_5/kgMLSS-d))} = 656 \ kg \ MLSS$$

$$Reactor \ Volume_{(low \ water \ level)} \ (m^3) = \frac{MLSS \ Mass \ (kg)}{MLSS \ Concentration \ (mg/L)} \times \frac{10^6 \ (mg/kg)}{10^3 \ (L/m^3)}$$

$$Reactor \ Volume_{(low \ water \ level)} = \frac{656 \ (kg)}{3\ 500 \ (mg/L)} \times 10^3 \ (mg-m^3/kg-L) = 188 \ m^3 \ (6\ 634 \ ft^3)$$

Since the decant volume represents 60% of the total volume,

Total Reactor Volume = $188/(1-0.6) = 470 \ m^3 \ (16\ 598 \ ft^3)$

*b. Decant Volume*

Total Decant Volume = Total Reactor Volume $(m^3)$ - Reactor Volume$_{(low \ water \ level)}$ $(m^3)$

Total Decant Volume = 470 $(m^3)$ - 188 $(m^3)$ = 282 $m^3$ (9 959 $ft^3$)

*c. Detention Time*

$$Max. \ Detention \ Time \ (hr) = \frac{Total \ Reactor \ Volume \ (m^3)}{Flow \ (L/d) \times 10^{-3} \ (m^3/L)} \times 24 \ (hr/d)$$

$$Max. \ Detention \ Time = \frac{470 \ (m^3) \times 24 \ (hr/d)}{378.5 \times 10^3 \ (L/d) \times 10^{-3} \ (m^3/L)} = 30 \ hr$$

$$Min. \ Detention \ Time \ (hr) = \frac{Decant \ Volume \ (m^3)}{Flow \ (L/d) \times 10^{-3} \ (m^3/L)} \times 24 \ (hr/d)$$

$$Min. \ Detention \ Time = \frac{282 \ (m^3) \times 24 \ (hr/d)}{378.5 \times 10^3 \ (L/d) \times 10^{-3} \ (m^3/L)} = 17.9 \ hr$$

*d. SBR Dimensions*

$$Basin \ Area \ (m^2) = \frac{Basin \ Volume_{low \ water \ level} \ (m^3)}{Minimum \ Depth \ (m)} = \frac{188 \ (m^3)}{2.75 \ (m)} = 68.4 \ m^2 \ (736 \ ft^2)$$

$$Basin \ Length = \sqrt{68.4 \ (m^2)} \approx 9 \ m \ (30 \ ft)$$

(Sheet 1 of 3)

Table E-10.  (Continued)

$$Basin\ Depth\ (m) = \frac{Total\ Reactor\ Volume\ (m^3)}{Basin\ Area\ (m^2)} = \frac{470\ (m^3)}{[9\ (m)]^2} = 5.8\ m\ (19\ ft)$$

e.  Aeration Power

$Nitrogeneous\ O_2\ Demand\ (kg\ O_2/d) = NH_3-N_{oxidized}\ (kg/d) \times kg\ O_2/kg\ BOD_5$

$NH_3-N_{oxidized}\ (kg/d) = TKN\ Removed\ (kg/d) - Synthesis\ N\ (kg/d)$

$TKN\ Removed = (40 - 5) \times 378.5 \times 10^{-3} = 13.25\ kg/d\ (29\ lb/d)$

Synthethis N = 5% waste activated sludge of total daily sludge production

$Sludge\ Production\ (kg/d) = Net\ Sludge\ Yield\ (kgMLSS/kgBOD_5) \times BOD_5\ Removed\ (kg/d)$

$Sludge\ Production = 0.76\ (kgMLSS/kgBOD_5) \times 85.2\ (kg/d) = 64.8\ kg/d\ (143\ lb/d)$

$Synthethis\ N = 0.05 \times 64.8\ (kg/d) = 3.24\ kg/d\ (7\ lb/d)$

$NH_3-N_{oxidized} = 13.25\ (kg/d) - 3.24\ (kg/d) = 10\ kg/d\ (22\ lb/d)$

$Nitrogeneous\ O_2\ Demand = 10\ (kgNH_3-N_{oxidized}/d) \times 4.6\ (kgO_2/kgNH_3-N_{oxidized}) = 46\ kgO_2/d$

$Carbonaceous\ O_2\ Demand\ (kg\ O_2/d) = BOD_5\ Mass\ (kg/d) \times kg\ O_2/kg\ BOD_5$

$Carbonaceous\ O_2\ Demand = 3.24\ (kgBOD_5/d) \times 1.28\ (kgO_2/kgBOD_5) = 4.15\ kgO_2/d$

AOR (kg/d) = Carbonaceous O2 Demand (kg/d) + Nitrogeneous O2 Demand (kg/d)

$AOR = 4.15\ kgO_2/d + 46\ kgO_2/d = 50.15\ kgO_2/d$

where:

AOR = Actual Oxygen Requirements (kg $O_2$/d)

$$SAOR\ (kg\ O_2/hr) = \frac{AOR \times C_S \times \Theta^{(T-20)}}{\alpha \times (\beta \times C_{SW} - C_O) \times Blower\ Usage\ (hr/d)}$$

where:

SAOR = Standard Actual Oxygen Requirements (kg $O_2$/d)
$\Theta$ (temperature correction factor) = 1.024
$C_S$ ($O_2$ saturation concentration at standard temperature and pressure) = 9.02 mg/L
$C_{SW}$ = concentration correction for elevation (i.e., 1 000 ft) = 9.02 - 0.0003 × elevation
$C_{SW}$ = 9.02 - 0.0003 × 1 000 = 8.72 mg/L (NOTE: 0.0003 may be used as a rule-of-thumb describing a 0.0003 mg/L rise/drop in DO saturation concentration per every foot of elevation increase/decrease.)
$C_O$ = 2 mg/L
$\alpha$ = 0.85; $\beta$ = 0.95; T = 20°C (67°F)
Blower Usage = 14 hr/d (based on 4 cycles per day (6 hr/cycle), 1.0 hr fill time, 3.5 hr react time, 0.75 hr settle time, 0.5 hr decant time, and 0.25 hr idle time)

(Sheet 2 of 3)

Table E-10. (Continued)

$$SAOR = \frac{50.15\ (kgO_2/d) \times 9.02\ (mg/L) \times 1.024^{(20\ -\ 20)}}{0.85 \times (0.95 \times 8.72\ -\ 2)\ (mg/L) \times 14\ (hr/d)} = 6.1\ kgO_2/hr\ (13.4\ lbO_2/hr)$$

$$Motor\ Requirements\ (kW) = \frac{SAOR\ (kgO_2/d)}{O_2\ Transfer\ Rate\ (kg/kW\text{-}d)}$$

$$Motor\ Requirements = \frac{6.1\ (kgO_2/hr)}{1.25\ (kgO_2/kW\text{-}hr)} = 4.9\ kW\ (6.5\ hp)$$

Since blowers typically have an efficiency of 50% or less, select two aerators with 11.2 kW (15 hp) motors. Blower size depends on the standard air flow rate. The standard air flow rate in standard cubic meters per minute (SCMM) is calculated as follows:

$$SCMM = \frac{\dfrac{SAOR\ (kgO_2/d)}{Blower\ Usage\ (hr/d)} \times \dfrac{1\ hr}{60\ min}}{O_2\ Content\ (kgO_2/m^3\ air) \times Absorption\ Efficiency}$$

where:
Air $O_2$ Content (at standard conditions) = 0.2793 kg $O_2/m^3$ of air
Obtain blower absorption efficiency from manufacturers

f. Sludge and Decant Flows

$$Sludge\ flow\ rate\ (L/d) = \frac{Sludge\ Mass\ Flow\ (kg/d)}{Sludge\ Density\ (kg/L)}$$

Typical sludge density = 1.02 kg/L

$$Decanter\ Flow\ Rate\ \ (L/min) = \frac{MDF}{NB \times NCB \times MCT}$$

where:
MDF = maximum daily flow for decant (or sludge waste)
NB　 = number of basins
NCB = number of cycles per basin
MCT = maximum cycle time for de cant or sludge waste (min)

(Sheet 3 of 3)

---

(2)　Blowers: rotary positive displacement.

(3)　Diffusers: 4-10 tube coarse bubble retriever diffuser assembly (2 per basin).

(4)　Mixers: 2 at 3.73 kW (5 hp).

(5)　Sludge pumps: 2 at 1.49 kW (2 hp).

(6)　Decanter sizing:　　　　　　cycles per day = 4.
volume per decant = 70.5 m$^3$ (2490 ft$^3$).
decant time = 30 min.
decant flow rate = 2.35 m$^3$/min (621 gal/min).

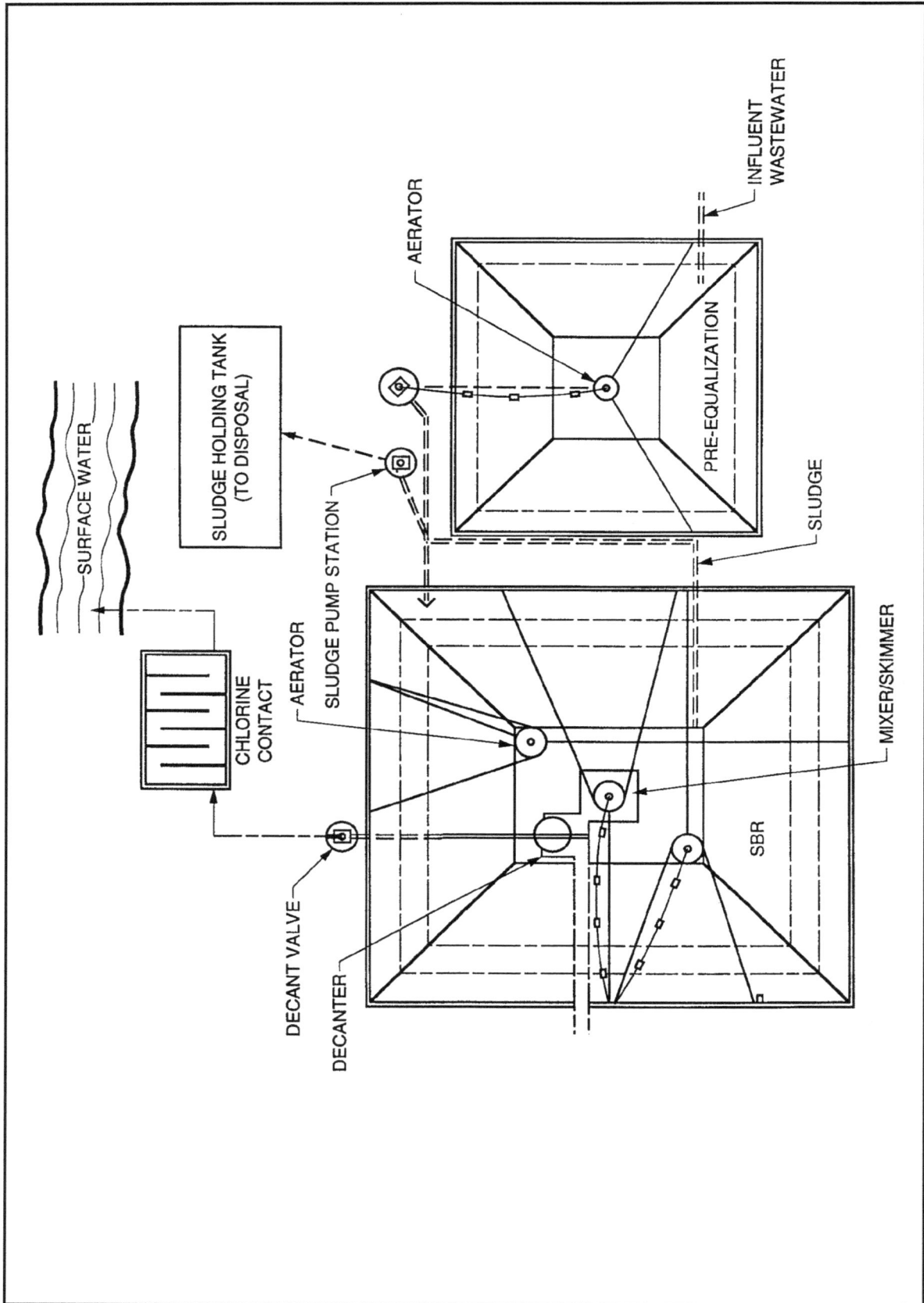

Figure E-4. Sequencing batch reactor

**Table E-11**
**Design Assumptions**

**Influent/Effluent Composition**

| Parameter | Influent | Effluent |
|---|---|---|
| $BOD_5$ | 250 mg/L | < 30 mg/L |
| TSS | 250 mg/L | < 30 mg/L |
| TKN | 20 mg/L | < 5 mg/L |
| Total Phosphorous | 10 mg/L | |
| Winter Temperature | 20°C (67°F) | |

**Assumptions**

| | | |
|---|---|---|
| L:W Ratio (for all basins): | 3:1 | |
| Hydraulic Loading: | 200-600 m³/ha-d | |
| | (20,000-65,000 gal/acre-d) | |
| $BOD_5$ Loading: | 40-80 kg/ha-d | |
| | (36-71 lb/acre-d) | |
| TKN Reduction: | 70-90% | |

(7)  Influent valves: 2, each 150 mm (6 in.) diameter.

(8)  Air blower values: 2, each 150 mm (6 in.) diameter.

## E-6. Constructed Wetlands Aerobic Non-Aerated Hyacinth System

Design an aerobic non-aerated hyacinth constructed wetlands secondary treatment system for a municipal wastewater flow of 284 000 L/d (75,000 gal/d).  The system will require preliminary or pretreatment (Imhoff tank).  Disinfection of effluent may be required depending upon regulatory restrictions.

*a.  BOD₅ loading.*  Calculate the influent $BOD_5$ loading using the influent $BOD_5$ concentration and the design flow, as shown in Table E-12.

*b.  Basin surface area.*  Calculate the required basin surface area at moderate $BOD_5$ loading rate of 50 kg/ha-d and the required area for the primary or first cells at a BOD5 loading rate of 100 kg/ha-d, as shown in Table E-12.

*c.  Primary cell dimensions.*  Use two primary cells.  Calculate the dimemsions of the two primary cells, as shown in Table E-12.

*d.  Final cell dimensions.*  Use four final cells.  Calculate the dimensions of the four final cells, as shown in Table E-12.

*e.  Cell volume.*  Calculate the primary and final cells volume, as shown in Table E-12.  Allow 0.5 m (1.6 ft) for sludge storage and assume 1.2 m (4 ft) effective water depth for a total pond depth of 1.7 m. Use a 3:1 sideslopes ratio to determine the treatment volume (approximation of a frustrum).

*f.  Hydraulic detention time.*  Estimate the hydraulic detention time in the effective, or above-sludge, level (zone) in the primary and final cells, as shown in Table E-12.

*g.  Hydraulic loading.*  Check the hydraulic loading to ensure that a minimum of 75 percent total nitrogen reduction is achieved to comply with the effluent quality requirements, as shown in Table E-12.

---

**Table E-12**
**Constructed Wetlands Aerobic Non-Aerated Hyacinth System**

---

*a. BOD$_5$ Loading*

$$BOD_5 \ Loading \ (kg/d) = Influent \ BOD_5 \ (mg/L) \times Flow \ (L/d) \times 10^{-6} \ (kg/mg)$$

$$BOD_5 \ Loading \ (kg/d) = 250 \ (mg/L) \times 284 \times 10^3 \ (L/d) \times 10^{-6} \ (kg/mg) = 71 \ kg/d \ (157 \ lb/d)$$

*b. Basin Surface Area*

$$Total \ Area \ Required \ (ha) = \frac{BOD_5 \ Loading \ (kg/d)}{Moderate \ Loading \ Rate \ (kg/ha-d)}$$

$$Total \ Area \ Required = \frac{71 \ (kg/d)}{50 \ (kg/ha-d)} = 1.42 \ ha \ (3.5 \ acre)$$

$$Area \ Primary \ Cells \ (ha) = \frac{BOD_5 \ Loading \ (kg/d)}{Primary \ Loading \ Rate \ (kg/ha-d)}$$

$$Area \ Primary \ Cells = \frac{71 \ (kg/d)}{100 \ (kg/ha-d)} = 0.71 \ ha \ (1.75 \ acre)$$

*c. Primary Cells Dimensions*

$$Cell \ Area \ (ha) = \frac{Area \ Primary \ Cells \ (ha)}{2} = L \times W = L \times \frac{L}{3}$$

$$Cell \ Area = \frac{0.71 \ (ha)}{2} \times 10 \ 000 \ (m^2/ha) = \frac{L^2}{3}$$

L = 103 m (338 ft)
W = 34 m (112 ft)

*d. Final Cells Dimensions*

Divide the remaining required area (1.42 ha - 0.71 ha = 0.71 ha) into two sets of two basins each (four cells of 0.18 ha each) to produce a total system with two parallel sets with three basins each.

$$Area \ Final \ Cells = 0.18 \ (ha) \times 10 \ 000 \ (m^2/ha) = \frac{L^2}{3}$$

L = 73.5 m (241 ft)
W = 24.5 m (80 ft)

*e. Cells Volume*

$$V = [(L)(W) + (L - 2sd)(W - 2sd) + 4(L - sd)(W - sd)] \times \frac{d}{6}$$

(Continued)

Table E-12 (Concluded)

where:
    V = basin volume ($m^3$)
    L = basin length at surface area (m)
    W = basin width at surface area (m)
    d = effective water depth = 1.2 m
    s = slope factor (e.g., 3:1 slope, s = 3)

V (primary cell) = 3631 $m^3$ (128,228 $ft^3$)

V (final cell) = 1758 $m^3$ (62,083 $ft^3$)

*f. Hydraulic Detention Time*

$$Detention\ Time\ (d) = \frac{No.\ Cells \times Unit\ Volume\ (m^3)}{Flow\ (m^3/d)}$$

$$Detention\ Time\ (Primary)\ t_p = \frac{2 \times 3\ 631\ (m^3)}{284\ (m^3/d)} = 26\ days$$

$$Detention\ Time\ (Final)\ t_p = \frac{4 \times 1\ 758\ (m^3)}{284\ (m^3/d)} = 25\ days$$

Total detention time = 26 + 25 = 51 days at 20° C (67°F)

*g. Hydraulic Loading*

$$Hydraulic\ Loading = \frac{284\ m^3/d}{1.42\ ha} = 200\ m^3/ha\text{-}d$$

This hydraulic loading is within the recommended range and is sufficient (i.e., $\leq$ 935 $m^3$/ha-d) to reduce the nitrogen loading by 70- 90 percent. It is reasonable to expect 5 mg/L of nitrogen or less in the effluent. Because the nitrogen will not be at optimum growth levels in this system, an annual harvest is suggested. An influent flow diffuser in each of the primary cells is recommended to properly distribute the untreated effluent.

# Appendix F
# U.S. Army Experience with Natural Wastewater Treatment Systems

*a. Background.* The most successful and probably the largest and longest operating U.S. Army Rapid Infiltration Basin installation is located at Fort Devens, Massachusetts (now transferred to the Commonwealth of Massachusetts under the Base Realignment and Closure plan). This facility was designed to treat 11 355 000 L/d (3,000,000 gal/d) of wastewater. The treatment system consists of a grit chamber with bypass and emergency overflow to the Nashua River; plant influent headworks (flow equalization tank, pumping station, bar racks, comminutors, wet well, Venturi flow meter and recorder), 3 Imhoff tanks, 22 rapid infiltration basins, and 4 sludge drying beds for the digested sludge from the Imhoff tanks.

*b. Performance.* Primary treatment of the wastewater is achieved in the Imhoff tanks. The effluent from the primary treatment system is evenly distributed over the 22 rapid infiltration basins with a total area of 18.5 ha (45.7 acre), and allowed to percolate. The percolated wastewater passes vertically downward through a 27 m (90 ft) soil column almost due east and some 122 m (400 ft) to the closest bank of the Nashua River. Soil characteristics of the underlying strata consist of generally poorly graded sands or gravelly sands with underlying lenses of silty sands and sandy gravels. Silt and clays probably constitute less than five percent of the underlying soils.

*c. Schematics.* A schematic flow diagram of the major unit processes is shown in Figure F-1. The schematic rapid infiltration basin and treated wastewater flow pattern is shown in Figure F-2.

*d. Fort Devens.* Except for relatively high nitrate ($NO_3^-$) concentrations detected in the fluctuating groundwater table (sometimes over 10 mg/L), the Fort Devens plant can be considered a successful natural treatment system. Table F-1 summarizes the $BOD_5$ loading and removal efficiency, and total nitrogen and phosphorous loadings and removal efficiencies from 1942 to 1990.

Table F-1
Wastewater Treatment Plant Performance, 1942-1990
Fort Devens, Massachusetts

|  | Pollutant Parameter | | |
|---|---|---|---|
|  | $BOD_5$ | Total Nitrogen | Phosphorous |
| Total applied (kg/ha-d) | 87 | 15 250 | -- |
| Total applied (mg/L) | 112 | 50 | 9.0 |
| Percolate (mg/L) | 12 | 10-20 | 0.10 |
| Applied BOD:N Ratio | -- | 2.4:1 | -- |
| Distance to sample point (m) | -- | -- | 45 |
| Percent Removal | 89% | 60-80% | 99% |

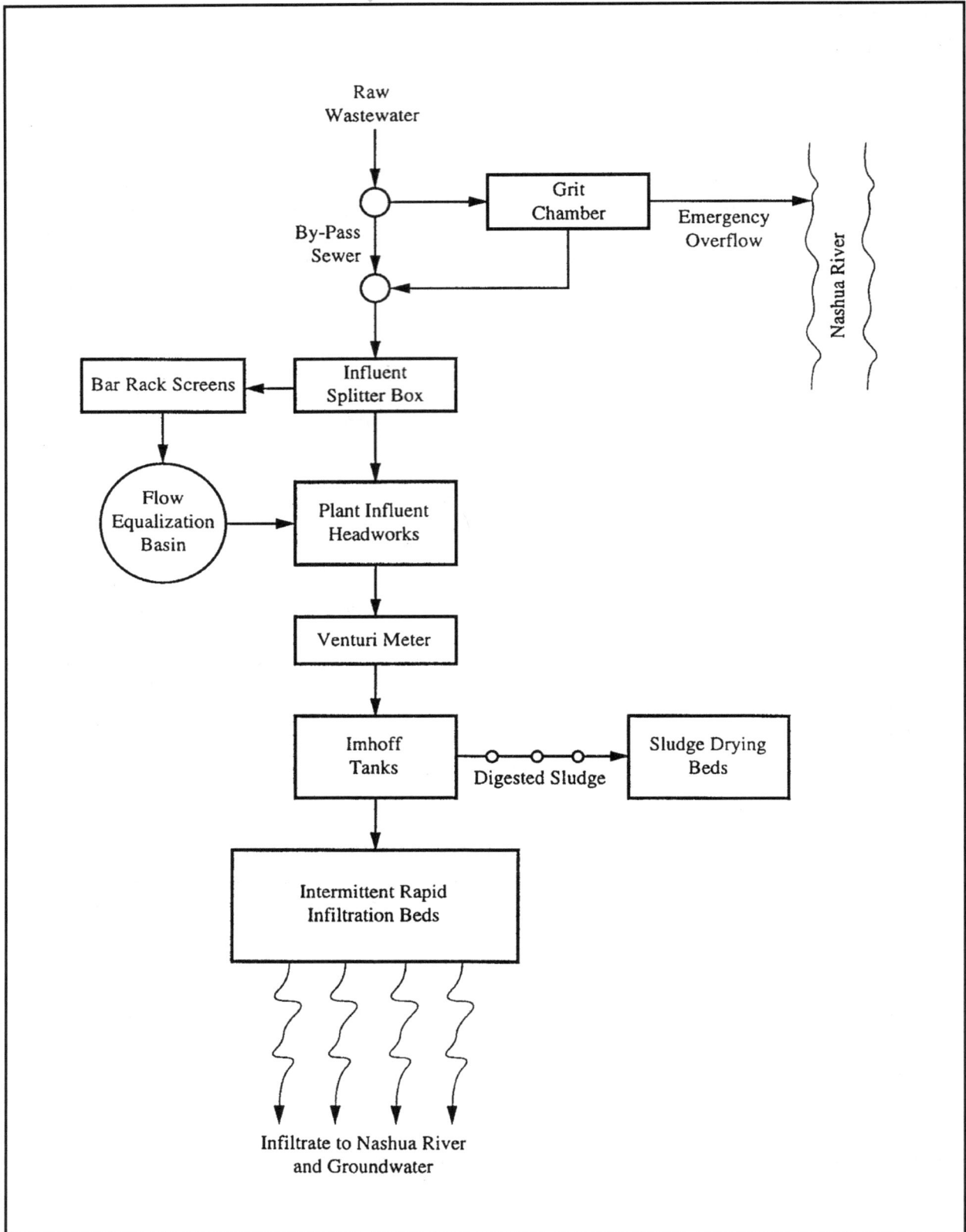

Figure F-1. Schematic flow diagram of Ft. Devens wastewater treatment facility

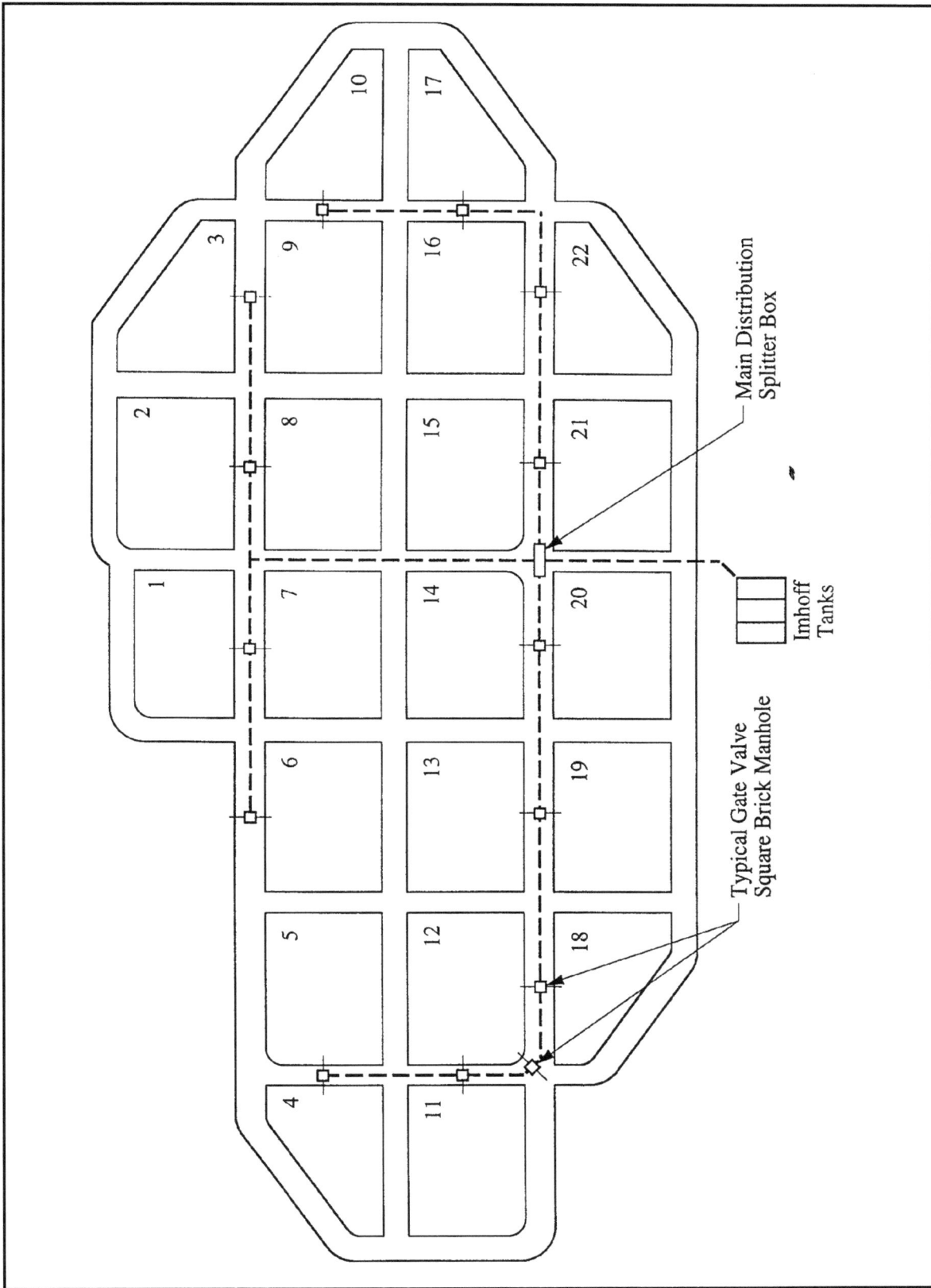

Figure F-2.  Ft. Devens schematic rapid infiltration basin plan and treated wastewater flow pattern

# Appendix G
# Abbreviations and Glossary of Terms

*Section I*
*Abbreviations*

**ANSI**
American National Standards Institute

**ATV**
Any Terrain Vehicle

**BPJ**
Best Professional Judgement

**CFR**
Code of Federal Regulations

**CLR**
Closed-Loop Reactor

**CWA**
Clean Water Act

**DNA**
Deoxyribonucleic acid

**ENRCC**
Engineering News-Record Construction Cost

**EPA**
Environmental Protection Agency

**FWS**
Free Water Surface

**MPN**
Most Probable Number

**MSC**
Major Subordinate Commands

**NPDES**
National Pollutant Discharge Elimination System

**NSFC**
National Small Flows Clearinghouse

**O&M**
Operation and Maintenance

**OF**
Overland Flow

**OSHA**
Occupational Safety and Health Administration

**PFRP**
Process to Further Reduce Pathogens

**POTW**
Publicly Owned Treatment Works

**PSRP**
Process to Significantly Reduce Pathogens

**RI**
Rapid Infiltration

**RBC**
Rotating Biological Contactor

**SBR**
Sequencing Batch Reactor

**SFS**
Subsurface Flow System

**SR**
Slow Rate

**TSS**
Total Suspended Solids

**USACE**
U.S. Army Corps of Engineers

**USACERL**
U.S. Army Construction Engineering Research Laboratory

**UV**
Ultraviolet

**WES**
Waterways Experiment Station

*Section II*
*Glossary of Terms*

**Absorption**
The taking up of one substance into the body of another.

**Acid**
(1) A substance that tends to lose a proton. (2) A substance that dissolves in water with the formation of hydrogen ions. (3) A substance containing hydrogen which may be replaced by metals to form salts.

**Acidity**
The quantitative capacity of aqueous solutions to react with hydroxyl ions. It is measured by titration with a standard solution of a base to a specified end point. Usually expressed as milligrams per liter of calcium carbonate.

**Activated Sludge**
Sludge floc produced in raw or settled wastewater by the growth of zoogleal bacteria and other organisms in the presence of dissolved oxygen and accumulated in sufficient concentration by returning floc previously formed.

**Activated Sludge Loading**
The pounds of biochemical oxygen demand (BOD) in the applied liquid-per-unit volume of aeration capacity or per pound of activated sludge per day.

**Activated Sludge Process**
A biological wastewater treatment process in which a mixture of wastewater and activated sludge is agitated and aerated. The activated sludge is subsequently separated from the treated wastewater (mixed liquor) by sedimentation and wasted or returned to the process as needed.

**Adsorption**
(1) The adherence of a gas, liquid, or dissolved material on the surface of a solid. (2) A change in concentration of gas or solute at the interface of a two-phase system. Should not be confused with absorption.

**Advanced Wastewater Treatment**
Those processes that achieve pollutant reductions by methods other than those used in conventional treatment (sedimentation, activated sludge, trickling filter, etc.). It employs a number of different unit operations, including lagoons, post-aeration, micro-straining, filtration, carbon adsorption, membrane solids separation, phosphorus removal, and nitrogen removal.

**Aerated Contact Bed**
A biological unit consisting of stone, cement-asbestos, or other surfaces supported in an aeration tank, in which air is diffused up and around the surfaces and settled wastewater flows throughout the tank. Also called contact aerator.

**Aerated Pond**
A natural or artificial wastewater treatment pond in which mechanical or diffused-air aeration is used to supplement the oxygen supply. See *oxidation pond*.

**Aeration**
The bringing about of intimate contact between air and a liquid by one or more of the following methods:
(a) spraying the liquid in the air; (b) bubbling air throughout the liquid; (c) agitating the liquid to promote
surface absorption of air. See following terms modifying *aeration:* diffused-air, mechanical, modified,
spiral-flow, step.

**Aeration Period**
(1) The theoretical time, usually expressed in hours, during which mixed liquor is subjected to aeration in
an aeration tank while undergoing activated sludge treatment. It is equal to the volume of the tank divided
by the volumetric rate of flow of the wastewater and return sludge. (2) The theoretical time during which
water is subjected to aeration.

**Aeration Tank**
A tank in which sludge, wastewater, or other liquid is aerated.

**Aerator**
A device that promotes aeration.

**Aerobic**
Requiring, or not destroyed by, the presence of free elemental oxygen.

**Aerobic Bacteria**
Bacteria that require free elemental oxygen for their growth.

**Aerobic Digestion**
Digestion of suspended organic matter by means of aeration. See *digestion.*

**Agglomeration**
The coalescence of dispersed suspended matter into larger flocs or particles which settle rapidly.

**Agitator**
(1) Mechanical apparatus for mixing and/or aerating. (2) A device for creating turbulence.

**Air**
The mixture of gases that surrounds the earth and forms its atmosphere, composed primarily of oxygen and
nitrogen. It also contains carbon dioxide, some water vapor, argon, and traces of other gases.

**Algae**
Primitive plants, one- or many-celled, usually aquatic, and capable of elaborating their foodstuffs by
photosynthesis.

**Alkali**
Any of certain soluble salts, principally sodium, potassium, magnesium, and calcium, that combine with
acids to form neutral salts and may be used in chemical processes such as water or wastewater treatment.

**Alkaline**
The condition of water, wastewater, or soil which contains a sufficient amount of alkali substances to raise
the pH above 7.0.

**Alkalinity**
The capacity of water to neutralize acids, a property imparted by the water's content of carbonates, bicarbonates, hydroxides, and occasionally borates, silicates, and phosphates. It is expressed in milligrams per liter of equivalent calcium carbonate.

**Alum**
A common name, in the water and wastewater treatment field, for commercial-grade aluminum sulfate.

**Aluminum Sulfate**
A chemical, sometimes called "waterworks alum" in water or wastewater treatment, prepared by combining bauxite with sulfuric acid.

**Ammonia**
A chemical combination of hydrogen (H) and nitrogen (N) occurring extensively in nature. The combination used in water and wastewater engineering is expressed in $NH_3$.

**Ammonia Stripping**
A modification of the aeration process for removing gases in water. Ammonium ions in wastewater exist in equilibrium with ammonia and hydrogen ions. As pH increases, the equilibrium shifts to the right, and above-pH-9 ammonia may be liberated as a gas by agitating the wastewater in the presence of air. This is usually done in a packed tower with an air blower.

**Ammonification**
Bacterial decomposition of organic nitrogen to ammonia.

**Anaerobic**
Requiring, or not destroyed by, the absence of air or free elemental oxygen.

**Anaerobic Bacteria**
Bacteria that grow only in the absence of free elemental oxygen.

**Anaerobic Contact Process**
An aerobic waste treatment process in which the microorganisms responsible for waste stabilization are removed from the treated effluent stream by sedimentation or other means and held in or returned to the process to enhance the rate of treatment.

**Anaerobic Denitrification**
A means to remove nitrates from wastewaters, especially irrigation return waters that may be high in nitrates and low in organics. In this method, an organic chemical such as methanol, ethanol, acetone, or acetic acid is added as a carbon source and the waste is placed in an anaerobic environment. Under these conditions, nitrate will be reduced by denitrifying bacteria to nitrogen gas and some nitrous oxide, which escapes to the atmosphere. With methanol, the chemistry can be represented as:

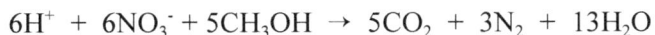

$$6H^+ + 6NO_3^- + 5CH_3OH \rightarrow 5CO_2 + 3N_2 + 13H_2O$$

**Anaerobic Digestion**
The degradation of organic matter brought about through the action of microorganisms in the absence of elemental oxygen.

**Anaerobic Digestion Process of Sewage Solids**
The first stage of the anaerobic digestion process of sewage solids is characterized by the production of organic acids. Proteins, carbohydrates, and fats are decomposed by the anaerobic bacteria, and the products of the decomposition are organic acids. This digestion stage is evident in sludge by a lowering of the pH and the presence of a disagreeable sour odor. Unless the amount of acid produced is excessive, the digestion will normally proceed to the second stage. With excess acidity, such as is obtained when the addition of fresh solids is too rapid, the bacteria will be destroyed and the process will end with the first stage. The second stage is characterized by liquefaction of sewage solids under mildly acid conditions. The bacteria, by enzyme action, convert the insoluble solids material to the soluble form. This is in accordance with the requirements of the bacterial cells that all food material must be in solution before it can pass through the cell wall. The third stage of digestion is characterized by production of gases, carbon dioxide, methane, and hydrogen sulfide, as well as an increase of pH and the production of carbonate salts.

**Anaerobic Waste Treatment**
Waste stabilization brought about through the action of microorganisms in the absence of air or elemental oxygen. Usually refers to waste treatment by methane fermentation.

**Boat**
Any vessel or other watercraft, privately owned or owned by the Corps of Engineers, whether moved by oars, paddles, sails, or other power mechanism, inboard or outboard, or any other vessel or structure floating on waters of a given state, whether or not capable of self-locomotion, including but not limited to cruisers, cabin cruisers, runabouts, houseboats, and barges. Excluded are commercial, passenger, or cargo-carrying vessels.

**Current**
(1) The flowing of water or other fluid. (2) That portion of a stream of water which is moving with a velocity much greater than the average or in which the progress of the water is principally concentrated.

**Cycle**
Filtration interval; length of time filter operates before cleaning.

**Dechlorination**
The partial or complete reduction of residual chlorine in a liquid by any chemical or physical process.

**Decomposition of Wastewater**
(1) The breakdown of organic matter in wastewater by bacterial action, either aerobic or anaerobic. (2) Transformation of organic or inorganic materials contained in wastewater through the action of chemical or biological processes.

**Defoamant**
A material having low compatibility with foam and a low surface tension. Defoamants are used to control, prevent, or destroy various types of foam, the most widely used being silicone defoamers. A droplet of silicone defoamant which contacts a bubble of foam will cause the bubble to undergo a local and drastic reduction in film strength, thereby breaking the film. Unchanged, the defoamant continues to contact other bubbles, thus breaking up the foam. A valuable property of most defoamants is their effectiveness in extremely low concentration. In addition to silicones, defoamants for special purposes are based on polymides, vegetable oils, and stearic acid.

**Defoaming Agent**
A material having low compatibility with foam and a low surface tension. See *defoamant*.

**Denitrification**
(1) Chemically bound oxygen in the form of either nitrates or nitrites is stripped away for use by microorganisms. This produces nitrogen gas which can bring up flow in the final sedimentation process. (2) An effective method of removing nitrogen from wastewater. (3) A biological process in which gaseous nitrogen is produced from nitrite and nitrate.

**Depth of Side Water**
The depth of a liquid measured along the inside of the vertical exterior wall of a tank.

**Detention Time**
The theoretical time required to displace the contents of a tank or unit at a given rate of discharge (volume divided by rate of discharge).

**Dewatering**
Any process of water removal or concentration of a sludge slurry, as by filtration, centrifugation, or drying. (A dewatering method is any process which will concentrate the sludge solids to at least 15 percent solids by weight.)

**Diatomaceous Earth**
A fine, siliceous earth consisting mainly of the skeletal remains of diatoms (unicellular organisms).

**Diatomaceous Earth Filter**
A filter used in water treatment in which a built-up layer of diatomaceous earth serves as the filtering medium.

**Diffused Air**
A technique by which air under pressure is forced into sewage in an aeration tank. The air is pumped down into the sewage through a pipe and escapes out through holes in the side of the pipe.

**Diffused Air Aeration**
Aeration produced in a liquid by air passed through a diffuser.

**Diffusion Aerator**
An aerator that blows air under low pressure through submerged porous plates, perforated pipes, or other devices so that small air bubbles rise through the water or wastewater continuously.

**Digested Sludge**
Sludge digested under either aerobic or anaerobic conditions until the volatile content has been reduced to the point at which the solids are relatively non-putrescible and inoffensive.

**Digestor**
A tank in which sludge is placed to permit digestion to occur. Also called sludge digestion tank. See *sludge digestion*.

**Digestion**
(1) The biological decomposition of organic matter in sludge, resulting in partial gasification, liquefaction, and mineralization. (2) The process carried out in a digester. See *sludge digestion*.

**Digestion Chamber**
A sludge-digestion tank. Frequently refers specifically to the lower or sludge-digestion compartment of an Imhoff tank.

**Digestion of Sludge**
Takes place in heated tanks where the material can decompose naturally and odors can be controlled.

**Digestion Tank**
A tank in which sludge is placed to permit digestion to occur. See *sludge digestion*.

**Diluent**
A diluting agent.

**Dilution**
Disposal of wastewater or treated effluent by discharging it into a stream or body of water.

**Discharge**
(1) As applied to a stream or conduit, the rate of flow or volume of water flowing in the stream or conduit at a given place and within a given period of time. (2) The passing of water or other liquid through an opening or along a conduit or channel. (3) The rate of flow of water, silt, or other mobile substance which emerges from an opening, pump, or turbine, or which passes along a conduit or channel, usually expressed as cubic feet per second, gallons per minute, or million gallons per day.

**Disinfectant**
A substance used for disinfection.

**Disinfected Wastewater**
Wastewater to which chlorine or other disinfecting agents have been added, during or after treatment, to destroy pathogenic organisms.

**Disinfection**
The treatment of the larger portion of microorganisms in or on a substance with the probability that all pathogenic bacteria are killed by the agent used.

**Disk Screen**
A screen in the form of a circular disk which rotates about a central axis perpendicular to its plane.

**Dissolved Air Flotation**
A process that adds energy in the form of air bubbles, which become attached to suspended sludge particles, increasing the buoyancy of the particles and producing more positive flotation.

**Dissolved Oxygen (DO)**
The oxygen dissolved in water, wastewater, or other liquid, usually expressed in milligrams per liter, parts per million, or percent of saturation.

**Dissolved Solids**
Theoretically, the anhydrous residues of the dissolved constituents in water. Actually, the term is defined by the method used in determination. In water and wastewater treatment the Standard Methods tests are used.

**Ditch**
A small artificial open channel or waterway constructed through earth or rock to convey water.

**Domestic Wastewater**
Wastewater derived principally from dwellings, business buildings, institutions, and the like. It may or may not contain groundwater, surface water, or storm water.

**Dose**
(1) The quantity of substance applied to a unit quantity of liquid for treatment purposes. It can be expressed in terms of either volume or weight, e.g., pounds per million gallons, parts per million, grains per gallon, milligrams per liter, or grams per cubic meter. (2) Generally, a quantity of material applied to obtain a specific effect.

**Drum Screen**
A screen in the form of a cylinder or truncated cone which rotates on its axis.

**Drying Bed**
A wastewater treatment unit usually containing a bed of sand on which sludge is placed to dry by evaporation and drainage.

**Dumping**
A method for solid waste disposal.

**Effective Size**
The diameter of the particles, spherical in shape, equal in size, and arranged in a given manner, of a hypothetical sample of granular material that would have the same transmission constant as the actual material under consideration.

**Efficiency**
(1) The relative results obtained in any operation in relation to the energy or effort required to achieve such results. (2) The ratio of the total output to the total input, expressed as a percentage.

**Effluent**
(1) A liquid which flows out of a containing space. (2) Wastewater or other liquid, partially or completely treated or in its natural state, flowing out of a reservoir, basin, treatment plant, or part thereof.

**Effluent Stream**
A stream or stretch of stream which receives water from groundwater in the zone of saturation. The water surface of such a stream stands at a lower level than the water table or piezometric surface of the groundwater body from which it receives water.

**Endogenous Respiration**
An auto-oxidation of cellular material, which takes place in the absence of assimilable organic material, to furnish energy required for the replacement of protoplasm.

**Environment**
The physical environment of the world consisting of the atmosphere, the hydrosphere, and the lithosphere.

**Environmental Pollution**
The presence of any foreign substance or interference (organic, inorganic, radiological, acoustic, or biological) in the environment (water, air, or land) which tends to degrade its quality so as to constitute a hazard or impair the usefulness of environmental resources.

**Equalization**
A process by which variations in flow and composition of a waste stream are averaged in an equalizing unit.

**Equalizing Basin**
A holding basin in which variations in flow and composition of a liquid are averaged. Also called balancing reservoir.

**Eutrophication**
(1) The normally slow aging process by which a lake evolves into marsh and ultimately becomes completely filled with detritus and disappears. (2) The intentional or unintentional enrichment of water.

**Evaporation**
(1) The process by which water becomes a vapor at a temperature below the boiling point. (2) The quantity of water that is evaporated; the rate is expressed in depth of water, measured as liquid water, removed from a specified surface per unit of time, generally in inches or centimeters per day, month, or year.

**Evaporation Rate**
The quantity of water, expressed in terms of depth of liquid water, evaporated from a given water surface per unit of time. It is usually expressed in inches depth per day, month, or year.

**Evapotranspiration**
Water withdrawn from soil by evaporation and/or plant transpiration. Considered synonymous with consumptive use.

**Evapotranspiration Potential**
Water loss that would occur if there was never was a deficiency of water in the soil for use by vegetation.

**Evapotranspiration Tank**
A tank, filled with soil and provided with a water supply, in which representative plants are grown to determine the amount of water transpired and evaporated from the soil under observed climatic conditions. Sometimes improperly referred to as a lysimeter.

**Excess Sludge**
The sludge produced in an activated sludge treatment plant that is not needed to maintain the process and is withdrawn from circulation.

**Extended Aeration**
A modification of the activated sludge process which provides for aerobic sludge digestion within the aeration system. The concept envisages the stabilization of organic matter under aerobic conditions, and disposal of the end products into the air as gases, with the plant effluent in the form of finely divided suspended and soluble matter.

**Facultative Anaerobic Bacteria**
Bacteria which can adapt to growth in the presence, as well as in the absence, of oxygen. May be referred to as facultative bacteria.

**Ferric Chloride**
A chemical ($FeCl_3$) often used for sludge conditioning.

**Filter**
A device or structure for removing solid or colloidal material, usually of a type that cannot be removed by sedimentation, from water, wastewater, or other liquid. The liquid is passed through a filtering medium, usually a granular material but sometimes finely woven cloth, unglazed porcelain, or specially prepared paper. There are many types of filters used in water or wastewater treatment. See *trickling filter*.

**Filter Bed**
(1) A type of bank revetment consisting of layers of filtering medium of which the particles gradually increase in size from the bottom upward. Such a filter allows the groundwater to flow freely, but it prevents even the smallest soil particles from being washed out. (2) A tank for water filtration having a false bottom covered with sand, as a rapid sand filter. (3) A pond with sand bedding, as a sand filter or slow sand filter.

**Filter Cake**
The dewatered sludge discharged from the filter, containing 65 to 80 percent moisture, depending upon the type of sludge, the type of dewatering equipment, and the conditioning of the sludge.

**Filter Cloth**
A fabric stretched around the drum of a vacuum filter.

**Filtered Wastewater**
Wastewater that has passed through a mechanical filtering process but not through a trickling filter bed.

**Filter Efficiency**
The operating results from a filter as measured by various criteria such as percentage reduction in suspended matter, total solids, biochemical oxygen demand, bacteria, and color.

**Filtering Medium**
(1) Any material through which water, wastewater, or other liquid is passed for the purpose of purification, treatment, or conditioning. (2) A cloth or metal material of some appropriate design used to intercept sludge solids in sludge filtration.

**Filter Loading**
Organically, the pounds of biochemical oxygen demand (BOD) in the applied-liquid-per-unit-filter-bed area or volume per day. Hydraulically, the quantity of liquid applied per unit filter bed area or volume per day.

**Filter Rate**

The rate of application of material to some process involving filtration, for example, application of wastewater sludge to a vacuum filter, wastewater flow to a trickling filter, or water flow to a rapid sand filter.

**Filter Run**

(1) The interval between the cleaning and washing operations of a rapid sand filter. (2) The interval between the changes of the filter medium on a sludge-dewatering filter.

**Filter Underdrains**

A system of underdraining for collecting water that has passed through a sand filter or biological bed.

**Filter Wash**

The reversal of flow through a rapid sand filter to wash clogging material out of the filtering medium and reduce conditions causing loss of head. See *backwash*.

**Filtrate**

The process of passing a liquid through a filtering medium (which may consist of granular material such as sand, magnetite, or diatomaceous earth, finely woven cloth, unglazed porcelain, or specially prepared paper) for the removal of suspended or colloidal matter.

**Filtration Rate**

The rate of application of wastewater to a filter, usually expressed in million gallons per acre per day or gallons per minute per square foot.

**Final Effluent**

The effluent from the final treatment unit of a wastewater treatment plant.

**Final Sedimentation**

The separation of solids from wastewater in a final settling tank.

**Final Sedimentation tank**

A tank through which the effluent from a trickling filter or an aeration or contact-aeration tank is passed to remove the settleable solids. Also called final settling basin. See *sedimentation tank*.

**Final Settling Tank**

A tank through which the effluent from a trickling filter or an aeration or contact-aeration tank is passed to remove the settleable solids. Also called final settling basin. See *sedimentation tank*.

**Fine Screen**

A relative term, usually applied to screens with openings of less than 2.54 mm (1 in.), but in wastewater treatment often reserved for openings that may be 1.651 mm (1/16 in.).

**Five-day BOD (BOD$_5$)**

That part of oxygen demand associated with biochemical oxidation of carbonaceous, as distinct from nitrogenous, material. It is determined by allowing biochemical oxidation to proceed, under conditions specified in Standard Methods, for 5 days.

**Flash Dryer**
A device for vaporizing water from partly dewatered and finely divided sludge through contact with a current of hot gas or superheated vapor. It includes a squirrel-cage mill for separating the sludge cake into fine particles.

**Flash Mixer**
A device for quickly dispersing chemicals uniformly throughout a liquid.

**Floc**
Small gelatinous masses formed in a liquid by a reaction of a coagulant added thereto, through biochemical processes, or by agglomeration.

**Flocculating Tank**
A tank used for the formation of floc by the gentle agitation of liquid suspensions, with or without the aid of chemicals.

**Flocculation**
In water and wastewater treatment, the agglomeration of colloidal and finely divided suspended matter after coagulation by gentle stirring by either mechanical or hydraulic means. In biological wastewater treatment where coagulation is not used, agglomeration may be accomplished biologically.

**Flocculation Agent**
A coagulating substance which, when added to water, forms a flocculent precipitate which will entrain suspended matter and expedite sedimentation; examples are alum, ferrous sulfate, and lime.

**Flocculator**
(1) A mechanical device to enhance the formation of flow in a liquid. (2) An apparatus for the formation of flow in water and wastewater.

**Flotation**
The raising of suspended matter to the surface of the liquid in a tank as scum—by aeration, the evolution of gas, chemicals, electrolysis, heat, or bacterial decomposition—and the subsequent removal of the scum by skimming.

**Flow Rate**
The rate at which a substance is passed through a system.

**Flow Regulator**
A structure installed in a canal, conduit, or channel to control the flow of water or wastewater at intake or to control the water level in a canal, channel, or treatment unit.

**Flume**
(1) An open conduit of wood, masonry, or metal constructed on a grade and sometimes elevated. Sometimes called aqueduct. (2) A ravine or gorge with a stream running through it. (3) To transport in a flume, as logs.

**Foam**
(1) A collection of minute bubbles formed on the surface of a liquid by agitation, fermentation, etc. (2) The frothy substance composed of an aggregation of bubbles on the surface of liquids by violent agitation or by the admission of air bubbles to liquid containing surface-active materials, solid particles, or both.

**Foam Separation**
The planned frothing of wastewater or wastewater effluent as a means of removing excessive amounts of detergent materials, through the introduction of air in the form of fine bubbles. Also called foam fractionation.

**Food-to-Microorganism Ratio**
An aeration tank loading parameter.

**Gravity Filter**
A rapid sand filter of the open type, the operating level of which is placed near the hydraulic grade line of the influent and through which the water flows by gravity.

**Grease**
In wastewater, a group of substances including fats, waxes, free fatty acids, calcium and magnesium soaps, mineral oils, and certain other non-fatty materials. The type of solvent and method used for extraction should be stated for quantitation.

**Grinding**
A process for solid waste handling and disposal by which refuse is reduced to less than 50 mm (2 in.) by a shredder. Also called shredding.

**Grit**
The heavy suspended mineral matter present in water or wastewater, such as sand, gravel, or cinders.

**Grit Chamber**
A detention chamber or an enlargement of a sewer designed to reduce the velocity of flow of the liquid to permit the separation of mineral from organic solids by differential sedimentation.

**Grit Collector**
A device placed in a grit chamber to convey deposited grit to a point of collection.

**Groundwater**
Subsurface water occupying the saturation zone, from which wells and springs are fed. In a strict sense the term applies only to water below the water table. Also called phreatic water, or plerotic water.

**Gutter**
An artificially surfaced, and generally shallow, waterway provided at the margin of a roadway for surface drainage.

**Halogen**
Any one of the chemically related elements—fluorine, chlorine, bromine, iodine, and astatine.

**Hardness**
A characteristic of water—imparted by salts of calcium, magnesium, and iron such as bicarbonates, carbonates, sulfates, chlorides, and nitrates—that causes curdling and increased consumption of soap, deposition of scale in boilers, damage in some industrial processes, and sometimes objectionable taste. See *carbonate hardness*.

**Head Loss**
The loss in liquid pressure resulting from the passage of the solution through a pipe, a channel, or a treatment unit.

**Heavy Metals**
Metals that can be precipitated by hydrogen sulfide in acid solution—for example, lead, silver, gold, mercury, bismuth, or copper.

**High-Rate Digestion**
Accelerated anaerobic digestion resulting primarily from thorough mixing of digester contents. May be enhanced by thermophilic digestion.

**High-Rate Filter**
A trickling filter operated at a high average daily dosing rate, usually between 10 and 40 mgd/acre including any recirculation of effluent.

**Horizontal Flow Tank**
A tank or basin, with or without baffles, in which the direction of flow is horizontal.

**Humus Sludge**
(1) Sludge deposited in final or secondary settling tanks following trickling filters or contact beds. (2) Sludge resembling humus in appearance.

**Hydraulic Loading**
The flow (volume per unit time) applied to the surface area of the clarification or biological reactor units (where applicable).

**Hydraulic Loss**
The loss of head attributable to obstructions, friction, changes in velocity, and changes in the form of the conduit.

**Hydraulic Radius**
The right cross-sectional area of a stream of water divided by the length of that part of its periphery in contact with its containing conduit; the ratio of area to wetted perimeter. Also called hydraulic mean depth.

**Hydraulic Surface Loading Influent**
(1) The flow (volume per unit time) applied to a unit of surface area (square ft), applicable to trickling filter and filtration processes. (2) Wastewater or other liquid—raw or partially treated—flowing into a reservoir, basin, treatment process, or treatment plant.

**Impeller**
A rotating set of vanes designed to impel rotation of a mass of fluid.

**Impervious**
Not allowing, or allowing only with great difficulty, the movement of water; impermeable.

**Infiltrate**
(1) To filter into. (2) The penetration by a liquid or gas of the pores or interstices.

**Infiltration**
(1) The flow or movement of water through the interstices or pores of a soil or other porous medium. (2) The quantity of groundwater that leaks into a pipe through joints, porous walls, or breaks. (3) The entrance of water from the ground into a gallery. (4) The absorption of liquid by the soil, either as it falls as precipitation or from a stream flowing over the surface. See *percolation*.

**Influent**
Water, wastewater, or other liquid flowing into a reservoir, basin, or treatment plant, or any unit thereof.

**Inhibitory Toxicity**
Any demonstrable inhibitory action of a substance on the rate of general metabolism (including rate of reproduction) of living organisms.

**Inorganic Matter**
Chemical substances of mineral origin, or more correctly, not of basically carbon structure.

**Intake**
(1) The works or structures at the head of a conduit into which water is diverted. (2) The process or operation by which water is absorbed into the ground and added to the saturation zone.

**Interface**
(1) A stratum of water of varying thickness lying between the fresh water above and ocean water below in certain estuaries. (2) A boundary layer between two fluids such as liquid-liquid or liquid-gas.

**Intermediate Screen**
A screen, with openings from 6.35 to 38.1 mm (0.25 to 1.5 in.), which prepares the waste flow for passage through grit chambers, primary sedimentation tanks, and reciprocating pumps.

**Intermediate Treatment**
Wastewater treatment such as aeration or chemical treatment, supplementary to primary treatment.

**Irrigation**
The artificial application of water to lands to meet the water needs of growing plants not met by rainfall.

**Lagoon**
A pond containing raw or partially treated wastewater in which aerobic or anaerobic stabilization occurs.

**Land Disposal**
Disposal of wastewater onto land.

**Lime**
Any of a family of chemicals consisting essentially of calcium hydroxide made from limestone (calcite) which is composed almost wholly of calcium carbonate or a mixture of calcium and magnesium carbonate.

**Liquid**
A substance that flows freely. Characterized by free movement of the constituent molecules among themselves, but without the tendency to separate from one another characteristic of gases. Liquid and fluid are often used synonymously, but fluid has the broader significance, including both liquids and gases.

**Liquid Sludge**
Sludge containing sufficient water (ordinarily more than 85 percent) to permit flow by gravity or pumping.

**Liquor**
Water, wastewater, or any combination; commonly used to designate liquid phase when other phases are present.

**Load**
See following terms modifying *load*: *BOD*, *peak*, *pollutional*.

**Loading**
The time rate at which material is applied to a treatment device involving length, area, or volume, or other design factor.

**Marina**
Any installation operating under public ownership which provides dockage or moorage for boats (exclusive of paddle or rowboats) and provides through sale, rental, or fee basis any equipment, supply, or service (fuel, electricity, or water) for the convenience of the public or its leasers, renters, or users of its facilities.

**Marine Sanitation Device**
Any equipment, piping, or appurtenances such as holding tanks for installation onboard a boat and any process to treat wastewater.

**Mechanical Aeration**
(1) The mixing, by mechanical means, of wastewater and activated sludge in the aeration tank of the activated sludge process to bring fresh surfaces of liquid into contact with the atmosphere. (2) The introduction of atmospheric oxygen into a liquid by the mechanical action of paddle, paddle wheel, spray, or turbine mechanisms.

**Mechanical Aerator**
A mechanical device for the introduction of atmospheric oxygen into a liquid. See *mechanical aeration*.

**Mechanical Agitation**
The introduction of atmospheric oxygen into a liquid by the mechanical action of paddle, paddle wheel, spray, or turbine mechanisms. See *mechanical aeration*.

**Mechanically Cleaned Screen**
A screen equipped with a mechanical cleaning apparatus for removal of retained solids.

**Mesh Screen**
A screen composed of woven fabric of any of various materials.

**Methane Fermentation**
Fermentation resulting in conversion of organic matter into methane gas.

**Microbial Activity**
Chemical changes resulting from the metabolism of living organisms. Biochemical action.

**Microbial Film**
A gelatinous film of microbial growth attached to or spanning the interstices of a support medium. Also called biological slime.

**Microbiology**
Study of very small units of living matter and their processes.

**Micron**
Unit of length: $10^{-6}$ ($39 \times 10^{-6}$ in).

**Microorganism**
Minute organism, either plant or animal, invisible or barely visible to the naked eye.

**Milligrams per Liter (mg/L)**
A unit of the concentration of water or wastewater constituent. It is 0.001 grams of the constituent in 1000 mL of water. It has replaced the unit formerly used commonly, parts per million, to which it is approximately equivalent, in reporting the results of water and wastewater analysis.

**Minimum Flow**
The flow occurring in a stream during the driest period of the year. Also called low flow.

**Mixed Liquor**
A mixture of activated sludge and organic matter undergoing activated sludge treatment in the aeration tank.

**Mixed-Liquor Volatile Suspended Solids (MLVSS)**
The concentration of volatile suspended solids in an aeration basin. It is commonly assumed to equal the biological solids concentration in the basin.

**Mixing Basin**
(1) A basin or tank wherein agitation is applied to water, wastewater, or sludge to increase the dispersion rate of applied chemicals. (2) A tank used for general mixing purposes.

**Mixing Tank**
A tank designed to provide a through mixing of chemicals introduced into liquids or of two or more liquids of different characteristics.

**Modified Aeration**
A modification of the activated sludge process in which a shortened period of aeration is used with a reduced quantity of suspended solids in the mixed liquor.

**Moisture**
Condensed or diffused liquid, especially water.

**Moisture Content**
The quantity of water present in soil, wastewater sludge, industrial waste sludge, and screenings, usually expressed in percentage of wet weight.

**Municipal Waste**
The combined residential and commercial waste materials generated in a given municipal area.

**Natural Water**
Water as it occurs in its natural state, usually containing other solid, liquid, or gaseous materials in solution or suspension.

**Nitrification**
(1) The conversion of nitrogenous matter into nitrates by bacterial. (2) The treatment of a material with nitric acid.

**Nitrosomonas**
A genus of bacteria that oxidize ammonia to nitrite.

**Nonbiodegradable**
Incapable of being broken down into innocuous products by the actions of living beings (especially microorganisms).

**Nonpotable Water**
Water which is unsatisfactory for consumption.

**Nonsettleable Matter**
That suspended matter which does not settle or float to the surface of water in a period of 1 hr.

**Nonsettleable Solids**
Wastewater matter that will stay in suspension for an extended period of time. Such period may be arbitrarily taken for testing purposes as 1 hr. See *suspended solids*.

**Nutrient**
(1) Any substance assimilated by organisms which promotes growth and replacement of cellular constituents. (2) A chemical substance (an element or an inorganic compound, e.g., nitrogen or phosphate) absorbed by a green plant and used in organic synthesis.

**Odor Control**
(1) In water treatment, the elimination or reduction of odors in a water supply by aeration, algae elimination, super-chlorination, activated carbon treatment, and other methods. (2) In wastewater treatment, the prevention or reduction of objectionable odors by chlorination, aeration, or other processes or by masking with chemical aerosols.

**Organic Loading**
Pounds of BOD applied per day to a biological reactor.

**Organic Matter**
Chemical substances of animal or vegetable origin, or more correctly, of basically carbon structure, comprising compounds consisting of hydrocarbons and their derivatives.

**Organic Matter Degradation**
The conversion of organic matter to inorganic forms by biological action.

**Orthophosphate**
An acid or salt containing phosphorus as $PO_4$.

**Overflow**
(1) The excess water that overflows the ordinary limits such as the stream banks, the spillway crest, or the ordinary level of a container. (2) To cover or inundate with water or other fluid.

**Overflow Rate**
One of the criteria for the design of settling tanks in treatment plants; expressed in gallons per day per square foot of surface area in the settling tank.

**Overland Runoff**
Water flowing over the land surface before it reaches a definite stream channel or body of water.

**Oxidation**
The addition of oxygen to a compound. More generally, any reaction which involves the loss of electrons from an atom.

**Oxidation Ditch**
A modification of the activated sludge process or the aerated pond, in which the mixture under treatment is circulated in an endless ditch and aeration and circulation are produced by a mechanical device such as a Kessener brush.

**Oxidation Pond**
A basin used for retention of wastewater before final disposal, in which biological oxidation of organic material is effected by natural or artificially accelerated transfer of oxygen to the water from air.

**Oxidation Process**
Any method of wastewater treatment for the oxidation of the putrescible organic matter. The usual methods are biological filtration and the activated sludge process.

**Oxidation Rate**
The rate at which the organic matter in wastewater is stabilized.

**Oxidized Sludge**
The liquid and solid product of the wet air oxidation of wastewater sludge.

**Oxidized Wastewater**
Wastewater in which the organic matter has been stabilized.

**Oxygen Demand**

(1) The quantity of oxygen utilized in the biochemical oxidation of organic matter in a specified time, at a specified temperature, and under specified conditions. See *BOD*.

**Oxygen Saturation**

The maximum quantity of dissolved oxygen that liquid of given chemical characteristics, in equilibrium with the atmosphere, can contain at a given temperature and pressure.

**Ozone**

Oxygen in molecular form with three atoms of oxygen forming each molecule ($O_3$).

**Parshall flume**

A calibrated device developed by Parshall for measuring the flow of liquid in an open conduit. It consists essentially of a contracting length, a throat, and an expanding length. At the throat is a sill over which the flow passes at Belanger's critical depth. The upper and lower head need not be measured unless the sill is submerged more than about 67 percent.

**Particle**

Any dispersed matter, solid or liquid, in which the individual aggregates are larger than single small molecules (about 0.0002 mm in diameter), but smaller than about 500 mm (20 in.) in diameter.

**Particle Size**

(1) The size of liquid or solid particles expressed as the average or equivalent diameter. (2) The sizes of the two screens, either in the U.S. Sieve Series or the Tyler Series, between which the bulk of a carbon sample falls, e.g., 8 x 30 means most of the carbon passes a No. 8 screen but is retained on a No. 30 screen.

**Parts per Million (ppm)**

The number of weight or volume units of a minor constituent present with each one million units of the major constituent of a solution or mixture. Formerly used to express the results of most water and wastewater analyses, but more recently replaced by the ratio mg/L.

**Pathogens**

Pathogenic or disease-producing organisms.

**Peak Demand**

The maximum momentary load placed on a water or wastewater plant or pumping statio or on an electric generating plant or system. This is usually the maximum average load in 1 hr or less, but may be specified as instantaneous or with some other short time period.

**Peak Load**

(1) The maximum average load carried by an electric generating plant or system for a short time period such as 1 hr or less. See *peak*. (2) The maximum demand for water placed on a pumping station, treatment plant, or distribution system, expressed as a rate. (3) The maximum rate of flow of wastewater to a pumping station or treatment plant. Also called peak demand.

**Percolating Filter**

A type of trickling filter.

## Percolation
(1) The flow or trickling of a liquid downward through a contact or filtering medium. The liquid may or may not fill the pores of the medium. Also called filtration. (2) The movement or flow of water through the interstices or the pores of a soil or other porous medium.

## pH
The reciprocal of the logarithm of the hydrogen-ion concentration. The concentration is the weight of hydrogen ions, in grams, per liter of solution. Neutral water, for example, has a pH value of 7 and a hydrogen-ion concentration of 10-7.

## Phosphate
A salt or ester or phosphoric acid.

## Pipe Gallery
(1) Any conduit for pipe, usually of a size to allow a man to walk through. (2) A gallery provided in a treatment plant for the installation of the conduits and valves and for a passageway to provide access to them.

## Pit Privy
A privy placed directly over an excavation in the ground.

## Pollution
A condition created by the presence of harmful or objectionable material or water.

## Pollutional Load
(1) The quantity of material in a waste stream that requires treatment or exerts an adverse effect on the receiving system. (2) The quantity of material carried in a body of water that exerts a detrimental effect on some subsequent use of that water.

## Polyelectrolyte
Long-chained, ionic, high-molecular-weight, synthetic, water-soluble, organic coagulants. Also referred to as polymers.

## Porous
Having small passages; permeable by fluids.

## Postchlorination
The application of chlorine to water or wastewater subsequent to any treatment, including prechlorination.

## Potable Water
Water that does not contain objectional pollution, contamination, minerals, or infective agents and is considered satisfactory for domestic consumption.

## Preaeration
A preparatory treatment of wastewater consisting of aeration to remove gases, add oxygen, promote flotation of grease, and aid coagulation.

## Prechlorination
The application of chlorine to water or wastewater prior to any treatment.

**Precipitation**
(1) The total measurable supply of water received directly from clouds as rain, snow, hail, or sleet; usually expressed as depth in a day, month, or year, and designated as daily, monthly, or annual precipitation. (2) The process by which atmospheric moisture is discharged onto a land or water surface. (3) The phenomenon that occurs when a substance held in solution in a liquid passes out of solution into solid form.

**Preliminary Treatment**
(1) The conditioning of a waste at its source before discharge, to remove or to neutralize substances injurious to sewers and treatment processes or to effect a partial reduction in load on the treatment process. (2) In the treatment process, unit operations, such as screening and comminution, that prepare the liquor for subsequent major operations.

**Presettling**
The process of sedimentation applied to a liquid before subsequent treatment.

**Pressure Regulator**
A device for controlling pressure in a pipeline or pressurized tank, such as a pressure-regulating valve or a pump drive-speed controller.

**Primary Settling Tank**
The first settling tank for the removal of settleable soils through which wastewater is passed in a treatment works.

**Primary Sludge**
Sludge obtained from a primary settling tank.

**Primary Treatment**
(1) The first major (sometimes the only) treatment in a wastewater treatment works, usually sedimentation. (2) The removal of a substantial amount of suspended matter but little or no colloidal and dissolved matter.

**Privy**
A building, either portable or fixed directly to a pit or vault, equipped with seating and used for excretion of bodily wastes.

**Privy Vault**
A concrete or masonry vault that is provided with a cleanout opening and over which is placed a privy building containing seats.

**Proportional Weir**
A special type of weir in which the discharge through the weir is directly proportional to the head.

**Public Water Supply**
A water supply from which water is available to the people at large or to any considerable number of members of the public indiscriminately.

**Pump-out Facilities**
Any device, equipment, or method of removing wastewater from a marine sanitation device, including any holding tanks either portable, movable, or permanently installed, and any wastewater treatment method or disposable equipment used to treat, or ultimately dispose of, wastewater removed from boats.

**Pumping Station**
A station housing relatively large pumps and their accessories. Pump house is the usual term for shelters for small water pumps.

**Purification**
The removal of objectionable matter from water by natural or artificial methods.

**Putrefaction**
Biological decomposition of organic matter with the production of ill-smelling products associated with anaerobic conditions.

**Radiation**
The emission and propagation of energy through space or through a material medium; also, the energy so propagated.

**Rakings**
The screenings or trash removed from bar screens cleaned manually or by mechanical rakes.

**Rapid Filter**
A rapid sand filter or pressure filter.

**Rapid Sand Filter**
A filter for the purification of water, in which water that has been previously treated, usually by coagulation and sedimentation, is passed downward through a filtering medium. The medium consists of a layer of sand, prepared anthracite coal, or other suitable material, usually 24-30 in. thick, resting on a supporting bed of gravel or a porous medium such as carborundum. It is characterized by a rapid rate of filtration, commonly from two to three gallons per minute per square foot of filter area.

**Raw Sludge**
Settled sludge promptly removed from sedimentation tanks before decomposition has much advanced. Frequently referred to as undigested sludge.

**Raw Wastewater**
Wastewater before it receives any treatment.

**Receiving Body of Water**
A natural watercourse, lake, or ocean into which treated or untreated wastewater is discharged.

**Recycling**
An operation in which a substance is passed through the same series of processes, pipes, or vessels more than once.

**Retention**
That part of the precipitation falling on a drainage area which does not escape as surface stream flow, during a given period. It is the difference between total precipitation and total runoff during the period, and represents evaporation, transpiration, sub-surface leakage, infiltration, and, when short periods are considered, temporary surface or underground storage on the area.

**Returned Sludge**
Settled activated sludge returned to mix with incoming raw or primary settled wastewater.

**Rotary Distributor**
A movable distributor made up of horizontal arms that extend to the edge of the circular trickling filter bed, revolve about a central post, and distribute liquid over the bed through orifices in the arms. The jet action of the discharging liquid normally supplies the motive power.

**Runoff**
(1) That portion of the earth's available water supply that is transmitted through natural surface channels. (2) Total quantity of runoff water during a specified time. (3) In the general sense, that portion of the precipitation which is not absorbed by the deep strata, but finds its way into the streams after meeting the persistent demands of evapotranspiration, including interception and other losses. (4) The discharge of water in surface streams, usually expressed in inches depth on the drainage area, or as volume in such terms as cubic feet or acre-feet. (5) That part of the precipitation which runs off the surface of a drainage area and reaches a stream or other body of water or a drain or sewer.

**Sand Filter**
A filter in which sand is used as a filtering medium. Also see *rapid sand filter, slow sand filter*.

**Sanitary Facilities**
Bathrooms, toilets, closets or other enclosures where commodes, stools, water closets, lavatories, showers, urinals, sinks, or other such plumbing fixtures are installed.

**Scale**
An accumulation of solid material precipitated out of waters containing certain mineral salts in solution and formed on interior surfaces, such as those of pipelines, tanks, and boilers, under certain physical conditions. May also be formed from interaction of water with metallic pipe.

**Screen**
A device with openings, generally of uniform size, used to retain or remove suspended or floating solids in flowing water or wastewater and to prevent them from entering an intake or passing a given point in a conduit. The screening element may consist of parallel bars, rods, wires, grating, wire mesh, or perforated plate, and the openings may be of any shape, although they are usually circular or rectangular.

**Screening**
The removal of relatively coarse floating and suspended solids by straining through racks or screens.

**Screenings**
Material removed from liquids by screens.

**Screenings Dewatering**
The removal of a large part of the water content of waste screenings by draining or by mechanical means.

**Screenings Grinder**
A device for grinding, shredding, or macerating material removed from wastewater by screens.

**Screenings Shredder**
A device that disintegrates screenings.

**Screw-feed Pump**
A pump with either horizontal or vertical cylindrical casing, in which operates a runner with radial blades like those of a ship's propeller.

**Scum**
(1) The layer or film of extraneous or foreign matter that rises to the surface of a liquid and is formed there. (2) A residue deposited on a container or channel at the water surface. (3) A mass of solid matter that floats on the surface.

**Secondary Settling Tank**
A tank through which effluent from some prior treatment process flows for the purpose of removing settleable solids. See *sedimentation tank*.

**Secondary Wastewater Treatment**
The treatment of wastewater by biological methods after primary treatment by sedimentation.

**Sedimentation**
The process of subsidence and deposition of suspended matter carried by water, wastewater, or other liquids, by gravity. It is usually accomplished by reducing the velocity of the liquid below the point at which it can transport the suspended material. Also called settling. See *chemical precipitation*.

**Sedimentation Basin**
A basin or tank in which water or wastewater containing settleable solids is retained to remove by gravity a part of the suspended matter. Also called sedimentation tank, settling basing, settling tank.

**Sedimentation Tank**
A basin or tank in which water or wastewater containing settleable solids is retained to remove by gravity a part of the suspended matter. Also called sedimentation tank, settling basin, settling tank.

**Septicity**
A condition produced by growth of anaerobic organisms.

**Septicization**
In anaerobic decomposition, the process whereby intensive growths of bacteria with the enzymes secreted by them liquefy and gasify solid organic matter.

**Septic Sludge**
Sludge from a septic tank or partially digested sludge from an Imhoff tank or sludge-digestion tank.

**Septic Tank**
A settling tank in which settled sludge is in immediate contact with the wastewater flowing through the tank and the organic solids are decomposed by anaerobic bacterial action.

**Septic Wastewater**
Wastewater undergoing putrefaction under anaerobic conditions.

**Settleable Solids**
(1) That matter in wastewater which will not stay in suspension during a preselected settling period, such as 1 hr, but either settles to the bottom or floats to the top. (2) In the Imhoff cone test, the volume of matter that settles to the bottom of the cone in 1 hr.

**Settled Wastewater**
Wastewater from which most of the settleable solids have been removed by sedimentation. Also called clarified wastewater.

**Settling**
The process of subsidence and deposition of suspended matter carried by water, wastewater, or other liquids, by gravity. It is usually accomplished by reducing the velocity of the liquid below the point at which it can transport the suspended material. Also called sedimentation. See chemical precipitation.

**Settling Basin**
A basin or tank in which water or wastewater containing settleable solids is retained to remove by gravity a part of the suspended matter. Also called sedimentation basin, sedimentation tank, settling tank.

**Settling Solids**
Solids that are settling in sedimentation tanks or sedimentation chambers and other such tanks constructed for the purpose of removing this fraction of suspended solids. See *settleable solids*.

**Settling Tank**
A basin or bank in which water or wastewater containing settleables solids is retained to remove by gravity a part of the suspended matter. Also called sedimentation basin, sedimentation tank, settling basin.

**Settling Velocity**
The velocity at which subsidence and deposition of the settleable suspended solids in water and wastewater will occur.

**Sewage**
The spent water of a community. Term now being replaced in technical usage by preferable term "wastewater." See *wastewater*.

**Sewer**
A pipe or conduit that carries wastewater or drainage water.

**Sewerage Facilities**
Entire wastewater collection and disposal system including commodes, toilets, lavatories, showers, sinks, and all other plumbing fixtures which are connected to a collection system consisting of sewer pipe, conduit, holding tanks, pumps and all appurtenances, including the wastewater treatment or disposal system.

**Sewer Gas**
Gas evolved in sewers that results from the decomposition of the organic matter in the wastewater.

**Sharp-Crested Weir**
A weir having a crest, usually consisting of a thin plate (generally of metal), so sharp that the water in passing over it touches only a line.

**Short-Circuiting**
A hydraulic condition occurring in parts of a tank where the time of travel is less than the flowing-through time.

**Shredder**
A device for size reduction.

**Shredding**
A process for the treatment and handling of solid wastes. The refuse is reduced to particles having no greater dimension than 2 in. by a shredder. Also called grinding.

**Side Water Depth**
The depth of water measured along a vertical exterior wall.

**Skimming**
The process of removing floating grease or scum from the surface of wastewater in a tank.

**Skimmings**
Grease, solids, liquids, and scum skimmed from wastewater settling tanks.

**Skimming Tank**
A tank designed so that floating matter will rise and remain on the surface of the wastewater until removed, while the liquid discharges continuously under curtain walls or scum boards.

**Slimes**
Substances of viscous organic nature, usually formed from microbiological growth.

**Slow Sand Filter**
A filter for the purification of water in which water without previous treatment is passed downward through a filtering medium consisting of a layer of sand or other suitable material, usually finer than for a rapid sand filter and from 610 mm to 1 m (24 to 40 in.) thick.

**Sludge**
(1) The accumulated solids separated from liquids, such as water or wastewater, during processing, or deposits on bottoms of streams or other bodies of water. (2) The precipitate resulting from chemical treatment, coagulation, or sedimentation of water or wastewater.

**Sludge Bed**
An area comprising natural or artificial layers of porous material on which digested wastewater sludge is dried by drainage and evaporation. A sludge bed may be open to the atmosphere or covered, usually with a greenhouse-type superstructure. Also called sludge drying bed.

**Sludge Blanket**
Accumulation of sludge hydrodynamically suspended within an enclosed body of water or wastewater.

**Sludge Cake**
The sludge that has been dewatered by a treatment process to a moisture content of 60-85 percent, depending on type of sludge and manner of treatment.

**Sludge Circulation**
The overturning of sludge in sludge-digestion tanks by mechanical or hydraulic means or by use of gas recirculation to disperse scum layers and to promote digestion.

**Sludge Collector**
A mechanical device for scraping the sludge on the bottom of a settling tank to a sump from which it can be drawn.

**Sludge Concentration**
Any process of reducing the water content of sludge that leaves the sludge in a fluid condition.

**Sludge Conditioning**
Treatment of liquid sludge before dewatering to facilitate dewatering and enhance drain ability, usually by the addition of chemicals.

**Sludge Density Index**
The reciprocal of the sludge volume index multiplied by 100.

**Sludge Dewatering**
The process of removing a part of the water in sludge by any method such as draining, evaporation, pressing, vacuum filtration, centrifuging, exhausting, passing between rollers, acid flotation, or dissolved-air flotation with or without heat. It involves reducing from a liquid to a spadable condition rather than merely changing the density of the liquid (concentration) on the one hand or drying (as in a kiln) on the other.

**Sludge Digestion**
The process by which organic or volatile matter in sludge is gasified, liquefied, mineralized, or converted into more stable organic matter through the activities of either anaerobic or aerobic organisms.

**Sludge-Digestion Gas**
Gas resulting from the decomposition of organic matter in sludge removed from wastewater and placed in a tank to decompose under anaerobic conditions. Also see *sewage gas, sludge digestion.*

**Sludge-Digestion Tank**
A tank in which sludge is placed for the purpose of permitting digestion to occur. See *sludge digestion.*

**Sludge Dryer**
A device for removal of a large percentage of moisture from sludge or screenings by heat.

**Sludge Drying**
The process of removing a large percentage of moisture from sludge by drainage or evaporation by any method.

**Sludge Filter**
A device in which wet sludge, usually conditioned by a coagulant, is partly dewatered by vacuum or pressure.

**Sludge Foaming**
An increase in the gas in sludge in Imhoff and separate digestion tanks, causing large quantities of froth, scum, and sludge to rise and overflow from openings at or near the top of the tanks.

**Sludge Lagoon**
A basin used for the storage, digestion, or dewatering of sludge.

**Sludge Reaeration**
The continuous aeration of sludge after its initial aeration for the purpose of improving or maintaining its condition.

**Sludge Reduction**
The reduction in quantity and change in character of sludge as the result of digestion.

**Sludge Solids**
Dissolved and suspended solids in sludge.

**Sludge Thickener**
A tank or other equipment designed to concentrate wastewater sludges.

**Sludge Thickening**
The increase in solids concentration of sludge in sedimentation or digestion tank. See *sludge concentration*.

**Sludge Treatment**
The processing of wastewater sludges to render them innocuous. This may be done by aerobic or anaerobic digestion followed by drying on sand beds, filtering and incineration, filtering and drying, or wet air oxidation.

**Sludge Utilization**
The use of wastewater sludges as soil builders and fertilizer admixtures. Sludges produced by aerobic and anaerobic digestion and activated sludge are used for these purposes.

**Sludge Volume Index (SVI)**
The ratio of the volume in milliliters of sludge settled from a 1 L sample in 30 min to the concentration of mixed liquor in milligrams per liter.

**Slurry**
A thin watery mud, or any substance resembling it, such as a lime slurry.

**Sodium Carbonate**
A salt used in water treatment to increase the alkalinity or pH value of water or to neutralize acidity. Chemical symbol is $Na_2CO_3$. Also called soda ash.

**Solids Retention Time**
The average residence time of suspended soils in a biological waste treatment system, equal to the total weight of suspended solids in the system divided by the total weight of suspended solids leaving the system per unit of time (usually per day).

**Specific Gravity**
(1) The ratio of the weight of a solid or liquid particle, substance, or chemical solution to the weight of an equal volume of water. Water has a specific gravity of 1.000 at 4 °C (39 °F). Particulates in raw water may have a specific gravity of 1.005 to 2.5.

(2) The ratio of the weight of a particular gas to an equal volume of air at the same temperature and pressure (air has a specific gravity of 1.0). Chlorine gas has a specific gravity of 2.5.

**Spiral Air Flow Diffusion**
A method of diffusing air in an aeration tank of the activated sludge process where, by means of properly designed baffles and the proper location of diffusers, a spiral or helical movement is given to the air and the tank liquor.

**Spiral Flow Aeration**
A method of diffusing air in an aeration tank of the activated sludge process. See *spiral air-flow diffusion*.

**Spiral-flow Tank**
An aeration tank or channel in which a spiral or helicoidal motion is given to the liquid in its flow through the tank by the introduction of air through a line of diffusers placed on one side of the bottom of each channel, by longitudinally revolving paddles, or by other means.

**Spray Irrigation**
A method for disposing of some organic wastewaters by spraying them on land, usually from pipes equipped with spray nozzles. This has proved to be an effective way to dispose of wastes from the canning, meat-packing, and sulfite-pulp industries where suitable land is available.

**Stabilization**
(1) Maintenance at a relatively non-fluctuating level, quantity, flow, or condition. (2) In lime-soda water softening, any process that will minimize or eliminate scale-forming tendencies. (3) In waste treatment, a process used to equalize wastewater flow composition prior to regulated discharge.

**Stabilization Lagoon**
A shallow pond for storage of wastewater before discharge. Such lagoons may serve only to detain and equalize wastewater composition before regulated discharge to a stream, but often they are used for biological oxidation. See *stabilization pond*.

**Stabilization Pond**
A type of oxidation pond in which biological oxidation of organic matter is effected by natural or artificially accelerated transfer of oxygen to the water from air.

**Step Aeration**
A procedure for adding increments of settled wastewater along the line of flow in the aeration tanks of an activated sludge plant.

## Sterilization

The destruction of all living microorganisms, as pathogenic or saprophytic bacteria, vegetative forms, and spores.

## Sterilized Wastewater

An effluent from a wastewater treatment plant in which all microorganisms have been destroyed by sterilization.

## Subsoil

That portion of a normal soil profile underlying the surface. In humid climates it is lower in content of organic matter, lighter in color, usually of finer particles, of denser structure, and of lower fertility than the surface soil. Its depth and physical properties control to a considerable degree the movement of soil moisture. In arid climates there is less difference between surface and subsoil.

## Sump

(1) A tank or pit that receives drainage and stores it temporarily, and from which the drainage is pumped or ejected. (2) A tank or pit that receives liquids.

## Supernatant

The liquid standing above a sediment or precipitate

## Surface Evaporation

Evaporation from the surface of a body of water, moist soil, snow, or ice. See *evapotranspiration.*

## Surface Wash

(1) A supplementary method of washing the filtering medium of a rapid sand filter by applying water under pressure at or near the surface of the sand by means of a system of stationary or rotating jets. (2) The surface runoff draining into a ditch or drain.

## Suspended Solids

Solids that either float on the surface of, or are in suspension in, water, wastewater, or other liquids, and which are largely removable by laboratory filtering.

## Tank

Any artificial receptacle through which liquids pass or in which they are held in reserve or detained for any purpose.

## Temperature

(1) The thermal state of a substance with respect to its ability to communicate heat to its environment. (2) The measure of the thermal state on some arbitrarily chosen numerical scale.

## Tertiary Treatment

A method used to refine the effluents from secondary treatment systems or otherwise increase the removal of pollutants.

## Thickened Sludge

A sludge concentrated to a higher solids content by gentle mixing, gravimetric settling, centrifugation, or air flotation.

**Thickener, Sludge**
A type of sedimentation tank in which sludge is permitted to settle, usually equipped with scrapers traveling along or around the bottom of the tank to push the settled sludge to a sump.

**Thickening Tank**
A sedimentation tank for concentrated suspensions.

**Total Kjeldahl Nitrogen (TKN)**
The sum of free ammonia and of organic compounds which are converted to $(NH_4)_2SO_4$ under the conditions of digestion.

**Total Organic Carbon (TOC)**
A measure of the amount of organic material in a water sample expressed in milligrams of carbon per liter of solution.

**Toxin**
Poisonous compounds produced by the metabolic activity or death and disintegration of microorganisms.

**Transient Slips**
Temporary docking or mooring space which may be used for short periods of time, including overnight, days, or weeks but less than 30 days.

**Trash**
Floating debris that may be removed from reservoirs, combined sewers, and storm-water sewers by coarse racks.

**Trash Rack**
A grid or screen placed across a waterway to catch floating debris.

**Trash Screen**
A screen installed or constructed in a waterway to collect and prevent the passage of trash.

**Treated Sewage**
Wastewater that has received partial or complete treatment.

**Treatment**
See following terms modifying *treatment: anaerobic waste, biological wastewater, chemical, intermediate, ion-exchange, preliminary, primary, secondary wastewater, sludge, waste, wastewater, water.*

**Trickling Filter**
A treatment unit consisting of a material such as broken stone, clinkers, slate, slats, or brush, over which sewage is distributed and applied in drops, films, or spray, from troughs, drippers, moving distributors, or fixed nozzles, and through which it trickles to the underdrains, giving opportunity for the formation of zoological slimes which clarify and oxidize the sewage.

**Trickling Filter Humus**
The sludge removed from clarifiers following biological stabilization in trickling filter units.

**Trickling Filter Ponding**
A condition occurring when voids in filter media become clogged with excessive growth of organisms, preventing the free flow of the wastewater.

**Trickling Filter Process**
In wastewater treatment, a process in which the liquid from a primary clarifier is distributed on a bed of stones. As the wastewater trickles through to drains underneath, it comes in contact with slime on the stones, by which organic material in the water is oxidized and impurities are reduced.

**Turbidity**
(1) A condition in water or wastewater caused by the presence of suspended matter, resulting in the scattering and absorption of light rays. (2) A measure of fine suspended matter in liquids. (3) An analytical quantity usually reported in arbitrary turbidity units determined by measurements of light diffraction.

**Ultraviolet Radiation**
Light waves shorter than visible blue-violet waves of the spectrum, having wave lengths of less than 3 900 Å.

**Ultraviolet Rays**
Those invisible light rays beyond the violet of the spectrum.

**Underdrain**
A drain that carries away groundwater or the drainage from prepared beds to which water or wastewater has been applied.

**Underflow**
(1) The movement of water through a given cross section of permeable rock or earth, possible under the bed of a stream or a structure. (2) The flow of water under a structure.

**Undigested Sludge**
Settled sludge promptly removed from sedimentation tanks before decomposition has much advanced. Also called raw sludge.

**Vacuum Filter**
A filter consisting of a cylindrical drum mounted on a horizontal axis, covered with a filter cloth, and revolving with a partial submergence in liquid. A vacuum is maintained under the cloth for the larger part of a revolution to extract moisture. The cake is scraped off continuously.

**Velocity**
See *settling velocity*.

**Viscosity**
The cohesive force existing between particles of a fluid which causes the fluid to offer resistance to a relative sliding motion between particles.

**Volatile**
Capable of being evaporated at relatively low temperatures.

**Volatile Solids**
The quantity of solids in water, wastewater, or other liquids, lost on ignition of the dry solids at 600°C (1 112°F).

**Wash Water**
Water used to wash filter beds in a rapid sand filter.

**Wash Water Gutter**
A trough or gutter used to carry away the water that has washed the sand in a rapid sand filter. Also called wash-water trough.

**Wash Water Rate**
The rate at which wash water is applied to a rapid sand filter during the washing process. Usually expressed as the rise of water in the filter in inches per minute or gallons per minute per square foot.

**Waste**
Something that is superfluous or rejected; something that can no longer be used for its originally intended purpose.

**Wasted Sludge**
The portion of settled solids from the final clarifier that was removed from the wastewater treatment processes and transferred to the solids handling facilities for ultimate disposal.

**Waste(s)**
See following terms modifying *waste(s)*: *industrial, municipal.*

**Waste Treatment**
Any process to which wastewater or industrial waste is subjected to make it suitable for subsequent use.

**Wastewater**
The spent water or wastewater containing human excrement coming from toilets, bathrooms, commodes and holding tanks. From the standpoint of source, it may be a combination of the liquid and water-carried wastes from residences, commercial buildings, industrial plants, and institutions, together with any groundwater, surface water, and storm water that may be present. Also referred to as sewage.

**Wastewater Decomposition**
Transformations of organic or inorganic materials contained in wastewater through the action of chemical or biological processes. See *decomposition of wastewater.*

**Wastewater Disposal**
The act of disposing of wastewater by any method (not synonymous with wastewater treatment). Common methods of disposal are dispersion, dilution, broad irrigation, privy, cesspool.

**Wastewater Facilities**
The structures, equipment, and processes required to collect, carry away, and treat domestic and industrial wastes, and dispose of the effluent.

**Wastewater Lagoon**

An impoundment into which wastewater is discharged at a rate low enough to permit oxidation to occur without substantial nuisance.

**Wastewater Treatment**

Any process to which wastewater is subjected in order to remove or alter its objectional constituents and thus render it less offensive or dangerous. See *intermediate treatment, primary treatment*.

**Wastewater Treatment or Disposal Systems**

The device, process, or plant designed to treat wastewater and remove solids and other objectionable constituents which will permit discharge to another approved system, or an approved discharge to state waters, or disposal through an approved subsurface drain field or other acceptable method.

**Wastewater Treatment Works**

(1) An arrangement of devices and structures for treating wastewater, industrial wastes, and sludge. Sometimes used synonymously with waste treatment plant or wastewater treatment plant. (2) A water pollution control plant.

**Water**

A transparent, odorless, tasteless liquid, a compound of hydrogen and oxygen, $H_2O$, freezing at $0\,°C$ ($32\,°F$) and boiling at $100\,°C$ ($212\,°F$), which, in more or less impure state, constitutes rain, oceans, lakes, rivers, and other such bodies; it contains 11.188 percent hydrogen and 88.812 percent oxygen, by weight. It may exist as a solid, liquid, or gas and, as normally found in the lithosphere, hydrosphere, and atmosphere, may have other solid, gaseous, or liquid materials in solution or suspension.

**Waterborne Disease**

A disease caused by organisms or toxic substances carried by water; the most common such diseases are typhoid fever, Asiatic cholera, dysentery, and other intestinal disturbances.

**Water Closet**

A plumbing fixture, usually a toilet bowl, seat, and water tank, or valved pressure water connection, for carrying off excreta and liquid wastes to a drain pipe connected below, by the agency of flushing water.

**Water Conditioning**

Treatments, exclusive of disinfection, intended to produce a water free of taste, odor, and other undesirable qualities.

**Water Treatment**

The filtration or conditioning of water to render it acceptable for a specific use.

**Water Treatment Plant**

That portion of water treatment works intended specifically for water treatment; may include, among other operations, sedimentation, chemical coagulation, filtration, and chlorination. See *water treatment works*.

**Water Treatment Works**

A group or assemblage of processes, devices, and structures used for the treatment or conditioning of water.

**Weir**

(1) A diversion dam. (2) A device that has a crest and some side containment of known geometric shape, such as a V, trapezoid, or rectangle, and is used to measure flow of liquid. The liquid surface is exposed to the atmosphere. Flow is related to upstream height of water above the crest, to position of crest with respect to downstream water surface, and to geometry of the weir opening.

**Weir Loading**

In a solids-contact or sedimentation unit, the rate in gallons per minute per foot of weir length at which clarified or treated liquid is leaving the unit. See *overflow rate*.

**Wet Well**

A compartment in which a liquid is collected, and to which the suction pipe of a pump is connected.

**Zooglea**

A jelly-like matrix developed by bacteria, associated with growths in oxidizing beds.

**Zoogleal Matrix**

The flow formed primarily by slime-producing bacteria in the activated sludge process or in biological beds.